4th Biennial Conference of the Interdisciplinary Society for Quantitative Research in Music and Medicine (ISQRMM 2017)

Published in the Journal of the Interdisciplinary Society for Quantitative Research in Music and Medicine

Bridgeport, Connecticut, USA
5 August 2017

ISBN: 978-1-5108-7520-3

Printed from e-media with permission by:

Curran Associates, Inc.
57 Morehouse Lane
Red Hook, NY 12571

ISBN: 978-1-5108-7520-3

Printed by Curran Associates, Inc. (2018)

For permission requests, please contact ISQRMM
at the address below.

ISQRMM
c/o David Akombo, Ph.D.
Jackson State University Department of Music
P.O. Box 17055
Jackson, MS 39217

http://isqrmm.org

The *Journal of the Interdisciplinary Society for Research in Music and Medicine* is a peer-reviewed journal presenting scholarly research into music's connection and application in medicine. Articles are submitted from scholars who have presented at the society's biannual conference, as well as from the field. Inquiries regarding submission should be directed to contact@isqrmm.com

Additional copies of this publication are available from:

Curran Associates, Inc.
57 Morehouse Lane
Red Hook, NY 12571 USA
Phone: 845-758-0400
Fax: 845-758-2634
Email: curran@proceedings.com
Web: www.proceedings.com

President's Forward

The study of the beneficial effects of music in medicine is a discipline that has spurred an interest among scholars from both the humanities and biomedical disciplines since the 18th century. Researchers in the 20th century embraced these studies with scholarly and academic vigor. Those in the 21st century have continued on with this inquiry. The mission of the Proceedings *Journal of the Interdisciplinary Society for Quantitative Research in Music and Medicine* is to provide global readership of peer reviewed articles that explore the unanswered questions regarding the efficacies of all forms of music on the human body, mind and soul. The Journal disseminates ISQRMM's biennial conference proceedings as well as new, unpublished and independent research papers not published or presented at the conferences. The Journal welcomes qualitative research as well as quantitative research methodologies and encourages interdisciplinary formats. The Journal encourages the authors to utilize modes of research that are clearly defined, explained and relevant. The Journal also welcomes Master's or Doctoral theses in any related fields.

The topics of historical qualitative research, regarding music and healing, especially as revealed in its manifold and remarkable manifestations—throughout Ancient Greece until the dawn of the 18th century—has never been candidly and fairly examined. There is also a dearth of research spanning the global cultures which are worthy of examination. Presently, the only sound sources of information on the culturally diverse aspects of music as medicine are anecdotes, which provide eyewitness accounts of individual music healers. In addition to the quantitative studies in the field, the anecdotal repository of evidence, as recorded in treatises of healing rituals, add credence to the historical implications of music as a healing agent.

The *Journal of the Interdisciplinary Society for Quantitative Research in Music and Medicine* balances its readership and uses an eclectic model to select the research papers submitted to make the Journal both scholarly and informative for all readers regardless of specialization. We are proud of ISQRMM supporters, enthusiasts, and our global clientele of our dedicated team of scholars.

David O. Akombo, Ph.D.
Co-Founder and Board Member

Editor's Forward

The Journal of the Interdisciplinary Society for Quantitative Research in Music and Medicine has collected articles reflecting the diverse backgrounds of music being used as a healing art around the world, as well as the many crossing points of the disciplines of music and medicine. The ISQRMM reflects many fields joined together in the pursuit of quantitative research. This is a platform which accommodates an expanse of various disciplines under one goal - the furtherment of the ancient and living art of promoting health and wellness with music.

I am honored to have served as editor for this journal, and found both the depth and the diversity of the articles of our contributors to be fascinating. It is my hope that the contents of this journal will spark further academic and clinical research, and promote discussion and collaboration in the pursuit of healing with music.

Raymond Hong
Board Member

Contents

A Pilot Study: The Effects of Music Therapy Techniques on the Expressive Language Skills of Adults in English as a Second Language Classes

Dr. Roy Kennedy

Abstract

The purpose of this study was to investigate the effects of music therapy techniques on the English Speaking Checklist and Story Retelling Inventory scores of adults in English as a Second Language (ESL) classes. Ten adult subjects participated in the study, all of whom spoke Spanish as their native language. The subjects were randomly assigned to either an experimental group (n = 5) or a control group (n = 5), with both groups receiving instruction for 12 weeks. The investigators used a wide variety of music therapy techniques as the primary instructional method for the experimental group, while the control group received instruction through traditional teaching methods. Music therapy techniques used include singing, listening, moving, playing, and improvising. Two independent observers collected data prior to and following treatment utilizing the English Speaking Checklist and Story Retelling Inventory. While there were not enough subjects to yield statistically significant results, the data does indicate that the experimental group made a 45.8% mean increase in score from pre-test to post-test on the English Speaking Checklist, whereas the control group exhibited only a 7.4% mean increase in performance.

Introduction

The United States has one of the best free public education systems in the world, with the average child receiving 12 years of education (Fox, 2007). Over 80 percent of the children in the United States graduate from high school, with approximately 70 percent going on to attend college. In stark contrast, children in Mexico receive an average of five years of education, with only five percent attending college. Furthermore, ten percent of Mexican adults are fully illiterate. These startling figures reveal several imminent problems facing Mexican citizens today. First of all, Mexico must improve its education system if it is to remain a viable contender in an increasingly competitive global economy. Secondly, a command of the English language is also necessary if Mexican citizens are to survive in a world where English is not only the national language of their largest trading partner, but also the emerging language of choice for world commerce and the Internet.

Mexico has already made great strides toward addressing these problems by implementing the Enciclomedia program. This unique program was launched with the goal of creating a nationwide network of electronic blackboards, linking every school in the country to computers and the Internet. As a result of this innovative approach, teachers and students are now able to teach themselves a wide variety of subjects (including English) through interactive learning. For Mexican immigrants living in the United States, however, the most effective means of acquiring the English language is through English as a Second Language (ESL) instruction.

Stephen Krashen (1981) identifies five key hypotheses related to second language acquisition: (1) the acquisition-learning distinction hypothesis, (2) the natural order hypothesis, (3) the Monitor hypothesis, (4) the input hypothesis, and (5) the affective filter hypothesis.

According to the acquisition-learning distinction hypothesis, language acquisition is a subconscious process through which an individual naturally acquires a feel for a language and its rules. By contrast, language learning is a conscious process through which an individual learns to apply specific rules of grammar.

Krashen's natural order hypothesis claims that grammatical structures tend to be acquired in a predictable order, meaning that certain grammatical structures are acquired earlier in the language acquisition process, while others are typically acquired later.

Acquired language and learned language have different functions. Language that is acquired subconsciously influences everyday interactions and overall fluency. Language that is learned consciously, however, is used as an editor in situations where an individual has enough time to edit, is focused on form, and knows the rule. Krashen refers to this conscious editing process as the Monitor. His Monitor hypothesis states that there are three kinds of Monitor users: (1) Monitor Over-users, (2) Monitor Under-users, and (3) Optimal Monitor users. As the name implies, Monitor Over-users employ their Monitor too frequently, causing them to become overly concerned with grammatical structure and thus greatly inhibiting their ability to speak fluently. Monitor Under-users, on the other hand, either have not learned the necessary grammatical rules, or choose not to use them, and therefore revert to subconsciously acquired language. Finally, Optimal Monitor users know to apply the Monitor in such a way that is situationally appropriate and does not interfere with communication. Optimal Monitor users utilize their editing skills in writing or in planned speech, but do not allow their conscious knowledge of grammatical structure to negatively affect normal conversation.

According to Krashen's input hypothesis, input must be comprehendible in order for language acquisition to occur. Furthermore, it is essential that language input be slightly above the individual's current level of ability. It is also of great importance that the individual be given a "silent period," during which the individual may acquire competence within a language before being required to produce it. This "silent period" is of particular interest to music therapists, as music therapy provides individuals with an enjoyable means of participating in the language acquisition process, without requiring them to produce the language directly.

Finally, Krashen claims that motivation, self-confidence, and anxiety all have an effect on language acquisition. As stated in his affective filter hypothesis, these three factors are able to directly increase or decrease the extent to which input is comprehended and language is acquired. Music therapy naturally provides a non-threatening environment in which an individual may meet his or her therapeutic goals through the use of fun and engaging activities, which makes it an excellent means of providing second language instruction. In addition, music therapy is easily tailored to meet each individual's specific needs, allowing him or her to acquire language at their own pace while simultaneously boosting confidence.

Previous research (Kennedy & Scott, 2005; Schunk, 1999) shows that music therapy has a significant positive impact on the academic skills of elementary and middle school students in English as a Second Language (ESL) classes. In 1999, Schunk investigated the effects of singing paired with signing on the receptive vocabulary skills of elementary ESL students. The results indicated that students placed in the singing paired with signing treatment group yielded the highest gains in score, suggesting that there are significant advantages to using music therapy as a means of aiding language acquisition at the elementary school level. Further research conducted in 2005 by Kennedy and Scott investigated the effects of music therapy on the academic skills of middle school ESL students. They concluded that music therapy does in fact have a significant positive impact on students' English speaking and story re-telling skills. Continuing this research, the purpose of the current study was to investigate whether or not music therapy is also capable of having a significant positive impact on the academic skills of adults in ESL classes.

Method

Subjects

Ten adult subjects participated in the study, three male and seven female. The subjects varied in age from their early twenties to their late forties. All of the subjects were native Spanish speakers attending beginning-level English as a Second Language classes organized by Catholic Charities in the southeastern part of the United States. The subjects were randomly assigned to either an experimental group (n = 5) or a control group (n = 5). Both groups received 12 weeks of instruction, with the subjects attending four 1.5-hour sessions each week. The experimental group received music therapy as a means of instruction, while the control group received instruction through traditional teaching methods.

Assessment

Two independent observers collected data prior to and following treatment utilizing the English Speaking Checklist and Story Retelling Inventory. Both of these assessment tools are taken from Hurley and Tinajero's textbook, *Literacy Assessment of Second Language Learners* (2001). Although there is currently no reliability or validity data available for these tests, Hurley and Tinajero claim that the *English Speaking Checklist* and *Story Retelling Inventory* are a much more effective means of assessing English as a Second Language learners' knowledge than traditional standardized tests.

Both the English Speaking Checklist and the Story Retelling Inventory are designed to acquire performance data related to a language learner's overall synthesis of spoken language. In contrast, standardized tests are only able to measure the learner's knowledge of specific facts or concepts. In addition, the English Speaking Checklist and Story Retelling Inventory are both capable of assessing ESL students' knowledge without wasting valuable instructional time. The tests are structured so that an observer (independent or otherwise) may collect data at any point during the learning process, informally and unobtrusively. Standardized tests, however, not only require that instructional time be set aside for assessment, but also demand a more formal testing environment which may result in test anxiety and less than optimal performance.

The English Speaking Checklist contains 15 items, each of which is satisfied by the observer circling "always," "sometimes," or "never." All of the items on the checklist are behaviors that are easily observed during everyday instruction, related to various aspects such as fluency, diction, coherence, organization, and courtesy. For example, the checklist contains items such as "sticks to the topic," "builds support for the subject," "speaks clearly," "takes turns and waits to talk," and "speaks loudly enough for others to hear," among others.

The Story Retelling Inventory is slightly more focused in scope than the English Speaking Checklist, and features ten items associated with a Likert scale. The observer is asked to rate the language learner's proficiency for each item on a scale from one to four, with 1 representing "weak," 2 representing "average," 3 representing "very good," and 4 representing "superior." The ten inventory items not only measure story comprehension, but also specific grammar skills and vocabulary usage. Examples of inventory items measuring story comprehension include "comprehensively retold the story," "comprehended the story line and plot," "understood role of characters," "understood implied and stated meaning," and "comprehended authors' intentions." Other items assess the learner's usage of verbs, adverbs, adjectives, phrases, compound sentences, vocabulary, slang, and colloquial expressions.

Procedure

At the request of Catholic Charities, both the experimental and control groups used the *Out and About* (Hemmert & Kappra, 2002) textbook as a guideline for instruction. The control group used traditional reading, writing, and speaking exercises to drill vocabulary and grammatical concepts, while the experimental music therapy group used singing, playing instruments, listening, moving, and improvising. Specific topics covered include greetings, farewells, the alphabet, numbers, ordinal numbers, directions, addresses, phone numbers, days of the week, months of the year, dates, time, age, residences, jobs, family, money, food, clothing, colors, parts of the body, weather, seasons, and holidays. Grammatical concepts were also addressed, including parts of speech and verb conjugation in the past, present, and future tenses.

Singing was the primary means of instruction for the experimental group, as it incorporates elements related to both reading and speaking skills. The therapists and subjects sang the majority of the songs in English, then verbally processed and translated the lyrics into Spanish. This process was reversed on occasion, giving the subjects a chance to sing in Spanish and translate into English. Song repertoire was selected from a broad range of styles and sources, including popular and folk music, as well as Elise Carr Sumner's "¡Viva la Música!" (2000), Stacey Tipton's "Musical Spanish" (2000), and Carolyn Graham's "Jazz Chants" (1978). Another technique frequently used was the piggybacking of new lyrics onto existing popular, folk, or children's tunes in order to address specific academic concepts.

In addition to singing activities, each session also featured several rhythmic activities designed to bring order to and energize the group, providing a sense of cohesion among its members. During these activities, the subjects played numerous rhythm instruments including djembes, hand drums, cow bells, jingle bells, tambourines, guiros, maracas, and egg shakers. The subjects also played pitched instruments such as the Boomwhackers, tone chimes, and Orff instruments.

The therapists used a wide variety of movement to music activities to teach parts of the body, as well as directions. For example, the subjects sang and danced "Father Abraham" to practice differentiating between right and left, teach commands such as "turn around" and "sit down," and reinforce vocabulary such as arm, foot, and head. "La Raspa" was used to teach the subjects specific parts of the foot, including heel, sole, and toes. The therapists also used the "Macarena" to teach specific parts of the hand (palm, fingers, and thumb) and directions (up and down). Other movement to music activities used were popular dance recordings such as the "Electric Slide" and the "Chicken Dance."

As a supplement to those activities previously mentioned, the therapists also incorporated several activities to address non-academic goals. For example, each session included at least one improvisation activity, which allowed the subjects to improve their non-verbal creative and expressive skills. Each session also ended with a 5-minute relaxation session, during which the subjects were taught to use progressive muscle relaxation techniques. The primary purpose of these activities was not only to provide the subjects with a wide variety of musical experiences, but also to give every subject an opportunity for immediate success (regardless of musical talent or linguistic aptitude).

Results

English Speaking Checklist

The first research question posed during this study was as follows: will there be a statistically significant difference in the experimental and control groups' pre-test to post-test scores on the English Speaking Checklist? Unfortunately, the low number of subjects (n = 10) does not lend itself to statistical analysis. However, through a simple comparison of pre-test to post-test scores, one can easily see a vast difference in the progress made by each group.

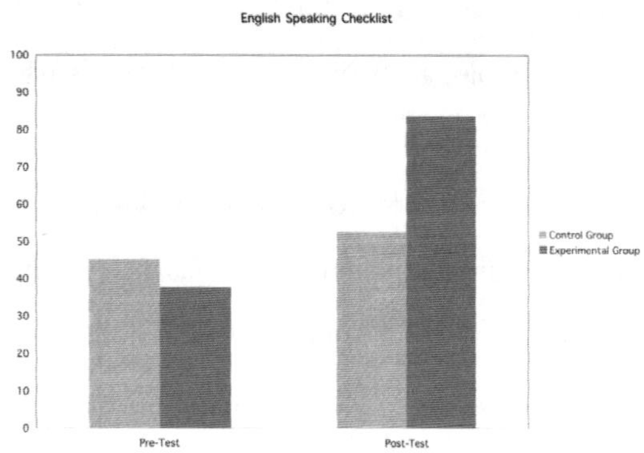

Figure 1

As seen in Figure 1, the mean pre-test score for the control group on the English Speaking Checklist was 45.2, while the mean pre-test score for the experimental group was 37.8. With regard to median pre-test scores, the control group's median score was 43, while the experimental group's median score was 33. Pre-test scores ranged from 27 to 67 for both the control and experimental groups. As can be seen, the control group scored 7.4% higher than the experimental group on the English Speaking Checklist pre-test. However, the control group scored 31% lower than the experimental group on the English Speaking Checklist post-test, with the mean for the control group being 52.6, and the mean for the experimental group being 83.6. The median post-test score for the control group was 53, with a median score of 90 for the experimental group. In terms of ranges, the control group's post-test scores on the English Speaking Checklist ranged from 43 to 63, while the experimental group's post-test scores ranged from 70 to 93. Figure 2 contains the pre-test to post-test scores of each subject in the experimental group on the English Speaking Checklist.

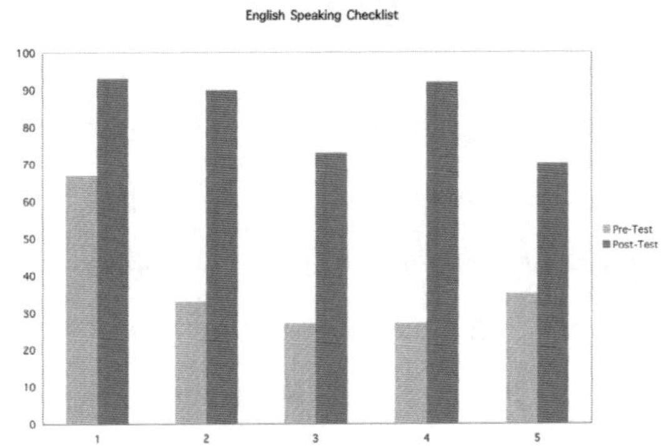

Figure 2

Story Retelling Inventory

The second research question posed by this study was the following: will there be a statistically significant difference in the experimental and control groups' pre-test to post-test scores on the Story Retelling Inventory? Due to the hectic schedules of the two independent observers, pre-test and post-test data from the Story Retelling Inventory was not able to be obtained for both the experimental and control groups. However, Figure 3 does display the pre-test to post-test scores for four of the five subjects in the experimental group. Post-test data was acquired for the fifth subject, although pre-test data was not.

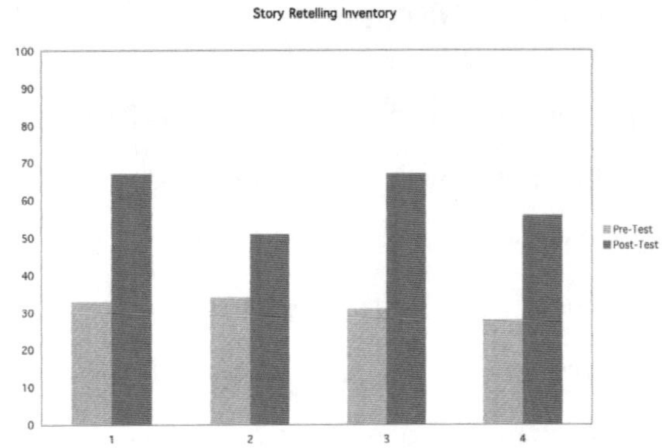

Figure 3

As can be seen in Figure 3, pre-test scores on the Story Retelling Inventory range from 28 to 34 for the experimental group, while post-test scores range from 51 to 72. The median pre-test score for the experimental group was 32, while the median score for the post-test was 67. Overall, there was a 31.1% increase from pre-test to post-test for the experimental group, with the mean pre-test score being 31.5, and the mean post-test score being 62.6.

Discussion

While the English Speaking Checklist and Story Retelling Inventory yielded a great deal of quantitative data for the experimental and control groups, the independent observers also recorded a significant amount of qualitative data by including additional comments and observations. These remarks cover a wide range of observable behaviors not included in the English Speaking Checklist or Story Retelling Inventory, including body language, affect, and the general tone of the class.

Many of the qualitative comments made refer to body language. For example, the independent observers noted such positive behaviors as “tapped pen in rhythm,” or “moving head back and forth to music.” Likewise, the observers also commented on some body language signs occurring during the control group that seem to indicate boredom, discomfort, or even frustration, including “scratches face, seemingly distracted, supporting face in hands,” “wrinkled forehead when asked questions by the teacher,” or “gave up and watched teacher.”

Other qualitative remarks pertain to the issue of affect. With regard to the experimental group, for example, the independent observers recorded such behaviors as “beamed at the therapist after she played xylophone,” “giggled during physical movement activity,” or “came ‘alive’ when the therapist played music.” Emotional states and affective behaviors observed during the control group include “hesitant,” “puzzled,” “frowned,” and “lost look.”

One of the greatest differences between the experimental and control groups addressed qualitatively was the issue of the class’ “tone” or “ambience.” One observer remarked on the control group’s pre-test that, “in my opinion, there was a lack of trust-building, community-building exercises here and less spontaneous joy in language use/class exercises.” The same observer later indicated that it was her “strong impression” that the “teacher needs to risk more, and students need to feel safer,” which directly relates to Stephen Krashen’s affective filter hypothesis previously mentioned. Another interesting comment made concerning the experimental group and pertaining to the affective filter hypothesis was that "while some teachers discourage native language use, this group of educators went back and forth between languages. Students were not 'divested' of the mother tongue." It was also noted of the experimental group that

> "what strikes me is the relaxed ambience in this class e.g. engaged students. It's difficult, in a way, to judge individual participation because the group seems cohesive. My guess is that the singing, dancing, and movement of teacher-performers and the students cuts down on fearfulness, performance anxiety, and some of the competition found in some conventional classes."

Perhaps one of the most striking comments, however, was as follows:

> "It occurs to me that this musical approach, which engages the mind and body on a 'primitive' level is good for mixed levels, including the inevitable range of abilities/levels in a single class. I say 'primitive' not as a pejorative: perhaps 'primal' or 'elemental' is better. Consider the bond an infant, toddler, or child makes with caretaker (mother or father). Optimally, there is pleasure involved in the relationship (touch, feeding, etc.). Ideally, there is pleasure involved in language learning."

In conclusion, while the data collected seems to indicate that music therapy does have a significant positive impact on the academic skills of adults in English as a Second Language classes, it is in no way intended to imply definite generalizability. The authors strongly recommend the replication of this study with a larger group of subjects so that statistically significant results may be obtained. Likewise, it is unfortunate that it was not possible to obtain control group data for the Story Retelling Inventory. Regardless of these minor details, however, it is still exciting to see the vast amount of improvement experienced by the experimental group as a result of music therapy. It is the authors' wish to see much more research conducted investigating the use of music therapy with this fascinating population.

References

Fox, V., & Allyn, R. (2007). *Revolution of hope: The life, faith, and dreams of a Mexican president*. Toronto, Canada: The Penguin Group.

Graham, C. (1978). *Jazz Chants*. New York, NY: Oxford University Press.

Hemmert, A., & Kappra, R. (2002). *Out and about: An interactive course in beginning English*. Burlingame, CA: Alta Book Center Publishers.

Hurley, S.L., & Tinajero, J.V. (2001). *Literacy assessment of second language learners*. Boston: Allyn and Bacon.

Kennedy, R., & Scott, A. (2005). A pilot study: The effects of music therapy interventions on middle school students' ESL skills. *Journal of Music Therapy, 42* (4), 244-261.

Krashen, S.D. (1981). *Principles and practice in second language acquisition*. London: Prentice Hall.

Schunk, H.A. (1999). The effect of singing paired with signing on receptive vocabulary skills of elementary ESL students. *Journal of Music Therapy, 36* (2), 110-124.

Sumner, E.C. (2000). *¡Viva la Música!* Anderson, IN.

Tipton, S. (2000). *Musical Spanish*. Winnipeg, Canada: The Musical Linguist.

A Small, Observational, Randomized Pilot Study to Test the Effects of Music on the Pain Perception of Sickle Cell Disease Patients

Kristie J. Lipford, Ph.D., Wendy Thompson, Ph.D., David O. Akombo, Ph.D.

Abstract

Introduction: The effects of music on pain have been examined, but very few studies have evaluated the influence of music therapy on sickle cell pain. To fill this gap, this research investigates the effects of music on the pain perception of sickle cell disease patients. Using data from a small pilot study, we examined the association of music-listening with pain perception among sickle cell patients.

Methods: Patients were recruited from the University of Mississippi Medical Center Sickle Cell Clinic during normal clinic hours in 2016. The study patient population consisted of 21 adult men and women who had a sickle cell disease diagnosis. The control group consisted of 9 patients and the experimental group included 12 patients. The analysis compared the self-assessed pain, number of provider phone calls, ER/hospital visits, medication usage, perceived benefit, and improvement in mood state among those participating in the home-based music-listening sessions to those of controls.

Results: Those in the control group were more likely to consume prescription and over-the-counter medication than those in the experimental group (.28 vs. .96). The experimental group reported fewer doctor calls (mean = 9.83 vs. 9.33), fewer doctor visits (mean = .96 vs. .95), and a lower level of average pain than the control group (mean = 4.71 vs. 5.11). Nearly half of the experimental group reported that the music-listening sessions were helpful in reducing pain (48.34%) and over a quarter described it as very helpful in mood improvement (28.34%).

Conclusions: In addition to music offering a complementary treatment method for those suffering from sickle cell derived pain, it also aids in stress reduction for sickle cell patients.

Introduction

Data reported by the National Heart, Lung, and Blood Institute (NHLBI) shows that there is an estimated 100,000 Americans who are affected with sickle cell disease (SCD) [1]. According to The Centers for Disease Control and Prevention, one out of every 500 African Americans are diagnosed with SCD. In addition, SCD occurs in nearly one out of every 365 Black or African-American births [2]. SCD is an inherited disorder of hemoglobin structure that has no established cure in adult patients. SCD is a quadfecta of (a) pain syndromes, (b) anemia and its sequelae, (c) organ failure, including infection, and (d) comorbid conditions [3]. Ultimately, pain episodes are a very common complication of SCD that unfortunately dominates the morbidity experience in SCD patients. The pathogenesis and management of acute and chronic sickle cell pain has been a concern to many SCD researchers for over a decade. Mitigating this pain can be a challenge since pains can be triggered differently ranging from those arising from the disease itself, those associated with therapy, and those that are due to comorbid conditions [3]. According to Ballas, the sequence of pathophysiologic events that lead to the perception of pain in SCD is not well known. It is agreed that tissue ischemia due to vascular occlusion resulting from in situ sickling causes infarctive tissue damage, which in turn initiates a secondary inflammatory response (p. 287). The many medications that are currently used to treat pain in SCD patients include antibiotics, hydroxyurea, gene therapy, nitric oxide, opioids, and statins. However, over the past few years, more patients are turning to complementary alternative medicine (CAM). CAM is a group of diverse medical and health care systems, practices, and products that are not generally considered part of conventional medicine [4]. Music-listening and music-making are considered to be forms of CAM that have been shown to be effective in pain management in a variety of medical conditions [5-11]. In a study on pain episodes and narcotic ingestion, Colwell (1997) found music to have positive benefits on medication intake [12]. Their findings showed that in randomized controlled trials (RTS), patients who listened to music consumed less prescription pain medication than those in the control group. This is an important finding because it suggests that music may be an effective route for non-opioid pain management.

Music has also been shown to beneficially affect stress-related physiological conditions [6, 8, 9, 13-15]. Research on the effects of music and music therapy for medical patients has burgeoned over the past 20 years, examining a variety of outcome measures in a wide range of specialty areas [13-16]. The literature suggests that increased anxiety leads to increased pain perception [17]. Recent studies show that music can lower both state and trait anxieties thereby helping SCD patients lower their pain perception [14]. This is especially important because prolonged experiences of stress and anxiety can lead to poor individual health as well as lower the quality of life. These can also trigger financial costs for families [18]. Therefore, developing cost-effective anxiety and stress management approaches, like the implementation of music therapy, has become an important endeavor of current research efforts, and thus the purpose of the current study.

Subjects

Patients were recruited from the University of Mississippi Medical Center (UMMC) Sickle Cell Clinic during normal clinic hours in spring 2016. The study patient population consisted of 21 adult men and women who had a sickle cell disease diagnosis. The control group consisted of 9 patients and the experimental group included 12 patients. Participants were included in the study if they were over 18 years of age, had a diagnosis of SCD, and had a history of SCD-derived pain. Patients were excluded if they had no English fluency or a mental handicap that prevented them from consenting and/or completing the procedures required for the study. This study was approved by Jackson State University Institutional Review Board and University of Mississippi Medical Center (UMMC) External Review Committee.

Measures

Outcome Measures

The participant's average daily pain level was also assessed for both the experimental and control group. We asked both groups, "Using the numeric pain rating scale, what was your average level of pain today?" The Numeric Pain Rating Scale (NPRS) is an 11 point (0-10) scale where higher scores indicate a higher level of pain [19]. The graphical scale depicts a ruler numbered from 0-10 indicating 0 as "no pain", 5 as "moderate pain", and 10 as "worst possible pain". This scale is widely used in patient care settings [20, 21]. An illustration of the scale was included in the Pain Diary for both the experimental and control group. Perceived pain for the experimental group was self-assessed and recorded before the music-listening session and immediately after the listening session using the NPRS. Pain medication usage was measured using two questions where higher response scores reflect the presence of no medication usage for both questions. Participants were asked, "Did you take any pain medication immediately *after* your music-listening session?" Responses for this question were "Yes" (0) or "No" (1). The second question asked participants to describe the type of medication consumed: "Was it prescription medication or over-the-counter medication (e.g. aspirin)?". Responses included, "Prescription" (0), "Over-the-Counter" (1), or "I didn't take any medication" (2).

Physician phone calls, physician visits, and ER visits were all patient-reported outcome measures. Participants were asked three questions regarding these behaviors, "Did you call your doctor's office or clinic today because of pain?", "Did you visit your doctor's office or clinic today because of pain?", and "Did you visit the ER (emergency room) today because of pain?". Responses for these questions were "Yes" (0) or "No" (1).

"Benefit in Reducing Pain" was measured using a single item. Participants were asked "How helpful was the music-listening in reducing your pain?" The responses choices included, "extremely helpful", "very helpful", "somewhat helpful", "slightly helpful", "not helpful at all". "Benefit in Mood Improvement" was assessed using one question. Participants were asked, "How helpful was the music-listening session in helping you to relax, feel better, uplift your mood, etc." Responses were on a 5-point Likert scale that ranged from "1 = not helpful at all" to "5 = extremely helpful" for both questions.

Additional Measures

Participants were also asked to record what genre or type of music they listened to and the length of their music-listening session. Both of these questions were open-ended. One question was asked to determine if participants used headphones or earbuds during their music-listening session.

Procedure

Over the span of 3 months, in 6 recruitment periods, 50 sickle cell disease patients met the inclusion criteria and were enrolled in this clinical study. Once patients agreed to participate in the study as indicated by a verbal consent and written consent (signing the informed consent form), we divided the patients into the clinical and the control group using a randomized assignment. Three patients, during the recruitment process, requested to be placed into the control group and this request was granted.

The research participants were provided pain journals at the start of the research project. The clinical group was asked to listen to a musical genre of their choice for 30 minutes each day for 30 days. Patients used their own music-listening devices and music. Patients without a portable listening device were provided mp3 players that included headphones. These mp3 players were loaded with 25-35 minutes of the musical genre that the patient stated they most preferred. These genres included rap/hip-hop, R&B, gospel, and jazz. Patients were instructed that the music could be listened to without headphones or with headphones, though we did express a study preference for music-listening with headphones or earbuds.

Specific instructions guided the participants' music-listening. When research participants in the experimental group experienced pain at a self-assessed level of 3 on the Numeric Pain Rating Scale (or 1 level before they generally take pain medication), they were instructed to listen to their preferred music for 30 minutes instead of taking prescribed medication. At the end of the 30-minute listening session, participants were asked to record, in a daily journal, an assessment of their perceived pain using the Numeric Pain Rating Scale. Participants were also asked to record their mood post-listening session, additional activities performed while music-listening, if they needed to take pain medication after their music-listening session, and if so, the type of pain medication (prescription opioids or over-the-counter pain killers). Those in the control group were required to record only five daily items, a) use of medication, b) physician phone call or c) physician visit due to SCD pain, d) ER visit because of SCD pain, and e) their average level of pain for that specific day.

Statistical Analysis

Descriptive statistics were used to determine the percentage differences between the experimental and control groups for medication usage type, doctor calls, doctor visits, emergency room visits, and average daily pain. Unpaired t-tests were used to determine the differences in pre- and post-pain level in music-listening. Independent samples t-tests were used to compare the differences between the experimental and control groups. We also investigated whether there were significant differences in pain levels, doctor calls, doctor visits, and ER visits by stratifying the sample by experimental and control groups. Descriptive statistics and t-tests were generated using SPSS 23.0.

Results

In total, there were 21 patients who were included in the data analysis. Although we recruited 50 patients, there were only 21 patients who were used for data analysis due to attrition and/or missing data. 12 patients were included in the experimental group and 9 patients comprised the control group. We also only included the first 5 days of the music-listening session due to extensive missing data as the study days progressed.

Average Minutes of Music-Listening and Most Preferred Music Genre

The average amount of music-listening for each day was 31.5 (SD=7.26). The most-preferred music genre was R&B for all 5 days, except for Day 4, where gospel was the most highly-listened category (33.3 percent for both categories).

Pain Perception

Average Daily Pain for Experimental and Control Group

We split the file to compare the average daily pain level of the clinical group and the experimental group. Results showed that the experimental group had a lower average of daily pain than the control group on 4 of the 5 days. Daily pain scores were summed, and an index was created to

Table 1. Mean scores of health behaviors by experimental vs. control group, 2016

	DAILY PAIN	MED USE	DR CALL	DR VISIT	ER
EXPERIMENTAL					
DAY 1	4.08 (3.14)	1.00 (1.04)	1.00 (.00)	1.00 (.00)	1.00 (.00)
DAY 2	4.16 (2.65)	.916 (.996)	1.00 (.00)	1.00 (.00)	1.00 (.00)
DAY 3	5.00 (2.66)	.916 (.900)	.91 (.28)	.91 (.28)	.91 (.28)
DAY 4	5.08 (2.71)	1.16 (1.02)	1.00 (.00)	.91 (.28)	.91 (.28)
DAY 5	5.25 (2.34)	.83 (1.02)	1.00 (.00)	1.00 (.00)	1.00 (.00)
CONTROL					
DAY 1	4.88 (3.05)	.22 (.44)	1.00 (.00)	1.00 (.00)	1.00 (.00)
DAY 2	4.33 (2.23)	.55 (.88)	1.00 (.00)	1.00 (.00)	1.00 (.00)
DAY 3	4.66 (3.74)	.22 (.66)	1.00 (.00)	1.00 (.00)	.88 (.33)
DAY 4	6.22 (2.12)	.00 (.00)	.77 (.44)	.88 (.33)	1.00 (.00)
DAY 5	5.33 (5.00)	.44 (.88)	.88 (.33)	.88 (.33)	1.00 (.00)
EXPERIMENTAL					
Sum Scores	23.58 (11.44)	4.88 (2.85)	4.91 (.28)	4.83 (.38)	4.83 (.38)
CONTROL					
Sum Scores	25.25 (10.27)	1.44 (1.42)	4.66 (.70)	4.77 (.66)	4.88 (.33)

determine a cumulative score of average pain for both the experimental and control group. These results show that the mean for the experimental group was 23.58 (SD=11.44, Mo=21) while the mean for the control group equaled 25.55 (SD=10.27, Mo=21). However, independent samples t-tests did not show any statistical significance on the average level of daily pain between the experimental and control group. Table 1 details these results.

Pre-Pain and Post-Pain for the Experimental Group

On Day 1, the average level of pre-pain was 4.75 (SD=2.83). This decreased by .92 with an average level of post-pain totaling 3.83 (SD=3.05). On Day 2, pre-pain levels were 4.50 (SD=2.57) while the post-pain totaled 3.75 (SD=2.56). The pre-pain mean for Day 3 was 5.33 (SD=2.49) with a decrease of 1.00 after the music-listening session (M=4.33, SD=2.67). On Day 4, the average level of pre-pain was 5.41 (SD=2.31). This decreased by 1.66 with an average level of post-pain totaling 3.75 (SD=2.59). On Day 5, the average level of pre-pain was 5.33 (SD=2.42). Pain levels were reduced by .67 after the listening session with the mean level of post-pain totaling 4.66 (SD=2.67).

Table 2: Mean levels of self-assessed pain pre- and post-music-listening session

	PRE-PAIN MEAN	POST-PAIN MEAN	SIG. LEVEL (2-tailed)
DAY 1	4.75 (SD=2.83)	3.83 (SD=2.24)	.009
DAY 2	4.50 (SD=2.57)	3.75 (SD=2.56)	.002
DAY 3	5.33 (SD=2.49)	4.33 (SD=2.67)	.015
DAY 4	5.41 (SD=2.31)	3.75 (SD=2.59)	.056
DAY 5	5.33 (SD=2.42)	4.66 (SD=2.67)	.005

T-test results of pre- and post-pain level N=12

A paired sample t-test was conducted to determine if the true mean difference is zero. Data revealed that there was a statistically significant difference between the mean pain level pre-treatment and post-treatment for all five days. Since the mean for pre-pain is greater than the mean for post-pain, we conclude that the music-listening session did have some effect on pain reduction for each day included in the analyses.

Physician phone calls, physician visits, and ER visits

The experimental group had fewer physician phone call and visits than the control group for the full five days of the music-listening session, as higher scores reflected no calls or physician and ER visits. Table 1 details means scores by each day while Table X shows the percentage of doctor calls and visits by each day.

Medication Usage

Overall, the experimental group had lower levels of medication usage than the control group. A sum of the five daily scores indicate that the control group's

Table X. Percentage scores of health behaviors by experimental vs. control group, 2016

	NO DR CALL	DR CALL	NO DR VISIT	DR VISIT	NO ER	ER
EXPERIMENTAL						
DAY 1	100	0	100	0	100	0
DAY 2	100	0	100	0	91.7	8.3
DAY 3	91.7	8.3	91.7	8.3	91.7	8.3
DAY 4	100	0	91.7	0	91.7	8.3
DAY 5	100	0	100	0	100	0
CONTROL						
DAY 1	100	0	100	0	100	0
DAY 2	100	0	100	0	100	0
DAY 3	100	0	100	0	88.9	11.1
DAY 4	77.8	22.2	88.9	11.1	100	0
DAY 5	88.9	11.1	88.9	11.1	100	0

mean was 1.44 (SD=1.42) while the experimental group's mean for medication usage was 4.83 (SD=2.85). Table 2 provides details of these daily mean scores. Table 3 describes the percentage scores and frequencies of medication use by medication type (prescription vs. over the counter). An independent samples t-test was performed comparing the summed mean scores of 5-day medication use for participants in the experimental and control group. Results indicate that the population variances significantly differ among these two groups (F =1.65, p < .005, two-tailed).

Table 3. Percentage scores of daily medication usage by experimental vs. control group, 2016

	PRESCRIPTION	OTC	NO MEDS
EXPERIMENTAL			
DAY 1	50.0	0	50
DAY 2	50.0	8.3	41.7
DAY 3	41.7	25.0	33.3
DAY 4	41.7	0	58.3
DAY 5	58.3	0	41.7
CONTROL			
DAY 1	77.8	22.2	0
DAY 2	66.7	11.1	22.2
DAY 3	88.9	11.1	0
DAY 4	100	0	0
DAY 5	77.8	0	22.2
EXPERIMENTAL			
Sum Scores	4.91 (.28)	4.83 (.38)	4.83 (.38)
CONTROL			
Sum Scores	4.66 (.70)	4.77 (.66)	4.88 (.33)

Perception of Helpfulness of Music in Reducing Pain and Mood Booster

There was a moderate agreement that the music-listening sessions were helpful. The overall mean for the total five days was 2.51 (SD=.43), which indicates that the majority of the experimental sample believed the music-listening to be slightly helpful (2.00) and somewhat helpful (3.00). Improvement in mood state was also assessed. The collective mean for the five days of music-listening was 3.25 (SD=.77) which suggests that the participants in the experimental group deemed the sessions to be somewhat helpful (3.00) and very helpful (4.00). Table 4 describes these results in greater detail.

Table 4: Percentage of perceived benefit of music-listening session

	NOT HELPFUL	SLIGHTLY HELPFUL	SOMEWHAT HELPFUL	VERY HELPFUL	EXTREMELY HELPFUL	MEAN SCORES
Perceived Benefit of Music in Reducing Pain						
DAY 1	8.3	25.0	50.0	16.7	0	2.75 (SD=.866)
DAY 2	0	0	66.7	33.3	0	2.66 (SD=.492)
DAY 3	8.3	41.7	41.7	8.3	0	2.50 (SD=.797)
DAY 4	16.7	25.0	50.0	8.3	0	2.50 (SD=.904)
DAY 5	16.7	50	33.3	0	0	2.16 (SD=.717)
Perceived Benefit of Music in Boosting Mood						
DAY 1	0	8.3	41.7	41.7	8.3	3.50 (SD=.797)
DAY 2	0	16.7	50.0	25.0	8.3	3.25 (SD=.866)
DAY 3	8.3	8.3	50	25.0	8.3	3.16 (SD=1.02)
DAY 4	8.3	25.0	25.0	25.0	16.7	3.16 (SD=1.26)
DAY 5	0	25.0	41.7	25.0	8.3	3.16 (SD=.937)

Percentage results of perceived benefit in reducing pain N=12

Discussion and Conclusion

This study examined the relationship between music-listening and pain perception among a sample of SCD patients. To our knowledge, no study has assessed this relationship with multiple days of music-listening. The data indicates that the music-listening session had some reduction on perceived pain level, which suggests that music-listening may be an integrative medical treatment for patients with sickle cell-derived pain. In addition to reducing pain, the majority of patients agreed that the music-listening session was helpful in de-stressing. This is an important finding because past studies have found that episodes of pain are preceded by high levels of stress, thus it may be helpful to implement stress reduction activities, like music-listening, in the treatment regimens of sickle cell patients. This study's findings can be explained by two brain mechanisms that may be responsible for the pain-relieving effect that music had on chronic pain in SCD patients who participated in the study. The first mechanism could be that enjoyable music selected by the SCD patients may have triggered the release of opioids in their brains. Opioids are the body's own 'morphine,' which may explain why music reduced the feeling of pain and reduced the need for pain medication. The second mechanism could be simply explained by attention deflection in that the pain-relieving effect may be the result of music simply being an incredibly effective way of redirecting their attention away from their pain, thereby applying the Gate Control Theory [22-24]. In this study, however, it appears most likely that the positive effect is due to the release of opioids in the brain, since some of the patients noted that the effect remained even after the music had stopped.

Despite these findings, this study has several limitations. The first limitation is the initial length of the study. We initially planned to collect patient data for 30 days, but this may have been an overburden to patients because there was missing data for a substantial number of participants after 5-7 days. The second limitation is that we relied on patients to collect data. Both the length of the study and reliance on self-reported patient data resulted in high attrition. The third limitation is that this was an uncontrolled experimental environment and we did not have much control over whether the participants actually listened to music, the length of time they listened to the music, and what activities they engaged in while listening to music. Finally, the fourth limitation is the limited demographic and baseline data we collected which restricted our ability to make comparisons of gender, SES, age, and time of SCD diagnoses.

Aside from these limitations, this study is noteworthy. It provides evidence for a non-pharmacological method to assist with reduction of SCD-associated pain as well as a mood booster. Furthermore, these results allow us to extend our understanding of music as an alternative form of medicine, which has larger implications for pain management.

References

1. National Heart, L., and Blood Institute. *What is sickle cell anemia?* What is sickle cell anemia? 2015 [cited 2017 February 9]; Available from: http://www.nhlbi.nih.gov/health/health-topics/topics/sca.
2. Prevention, C.f.D.C.a. *Data & Statistics Sickle Cell Disease (SCD).* Data & Statistics Sickle Cell Disease (SCD) 2016 [cited 2016 February 9]; Available from: https://www.cdc.gov/ncbddd/sicklecell/data.html.
3. Ballas, S.K., *Pain management of sickle cell disease.* Hematology/Oncology Clinics of North America, 2005. **19**: p. 785-802.
4. Medicine, N.C.f.C.a.A. *The use of complementary and alternative medicine in the United States.* The use of complementary and alternative medicine in the United States 2008 [cited 2017 February 9]; Available from: https://nccih.nih.gov/sites/nccam.nih.gov/files/camuse.pdf.
5. Gutgsell, K.J., et al., *Music therapy reduces pain in palliative care patients: a randomized controlled trial.* Journal of Pain and Symptom Management, 2013. **45**(5): p. 822-831.
6. Prensner, J.D., et al., *Music therapy for assistance with pain and anxiety management in burn treatment.* Journal of Burn Care and Research, 2001. **22**(1): p. 83-88.
7. Groen, K.M., *Pain assessment and management in end of life care: a survey of assessment and treatment practices of hospice music therapy and nursing professionals.* Journal of Music Therapy, 2007. **44**(2): p. 90-112.
8. Tan, X., et al., *The efficacy of music therapy protocols for decreasing pain, anxiety, and muscle tension levels during burn dressing changes: a prospective randomized crossover trial.* Journal of Burn Care and Research, 2010. **31**(4): p. 590-597.
9. Nilsson, U., *The anxiety-and pain-reducing effects of music intervemtions: a systematic review.* AORN journal, 2008. **87**(4): p. 780-807.
10. Parker, D.B., *Effect of music therapy for pain and anxiety versus literature on the immediate and future perceptions of cardiac patients*, in *School of Music*. 2017, Florida State University: Tallahassee, FL.
11. Lee, J.H., *The effects of music on pain: a meta-analysis.* Journal of Music Therapy, 2016. **53**(4): p. 430-477.
12. Colwell, C.M., *Music as distraction and relaxation to reduce chronic pain and narcotic ingestion: A case study.* Music Therapy Perspectives, 1997. **15**(1): p. 24-31.
13. Blood, A. and R. Zatorre, *Intensely pleasurable responses to music correlate with activity in brain regions implicated in reward and emotion.* Proceedings of the National Academy of Science USA, 2001. **98**(20): p. 11818-11823.
14. Akombo, D., *Effects of listening to music as an intervention for pain and anxiety in bone marrow transplant patients.* Dissertation Abstracts International, 2007. **68**(1-A).

15. Dileo, C. and J. Bradt, *Medical music therapy: a meta-analysis & agenda for future research*. 2005, Cherry Hill, NJ: Jeffrey Books.
16. Beck, S., *The effect of therapeutic use of music on cancer related pain* Oncology Nursing Forum 1991. **18**(8): p. 1327-1337.
17. McCracken, L. and E. Keogh, *Acceptance, mindfulness, and values-based action may counteract fear and avoidance of emotions in chronic pain: an analysis of anxiety sensitivity.* Journal of Pain, 2009. **10**(4): p. 408-415.
18. McEwen, B., *Protective and damaging effects of stress mediators.* New England Journal of Medicine, 1998. **338**(171-179).
19. McCaffery, M. and A. Beebe, *Pain: Clinical manual for nursing practice*. 1989, St. Louis, MO: Mosby-Year Book.
20. Williamson, A. and B. Hoggart, *Pain: a review of three commonly used pain rating scales.* Journal of Clinical Nursing, 2005. **14**(7): p. 798-804.
21. Ferreira-Valente, M.A., J.L. Pais-Ribeiro, and M.P. Jensen, *Validity of four pain intensity rating scales.* Pain, 2011. **152**(10): p. 2399-2404.
22. Melzack, R. and P.D. Wall, *Pain mechanisms: a new theory* Survey of Anesthesiology, 1967. **11**(2): p. 89-90.
23. Fellers, J.C., ed. *Theories of pain and addiction: type of pain, pathways to opiate addiction*. Treating comorbid opioid use disorder in chronic pain, ed. A. Matthews and J.C. Fellers. 2016, Springer International Publishing: Gewerbestrasse, Switzerland. 3-12.
24. Melzack, R. and P. Wall, *Pain mechanism: a new theory.* Science, 1965. **150**(3699): p. 971-979.

Music and Health in Africa: The Case for Drumming and Post Traumatic Stress Disorder

David O. Akombo
Jackson State University

Abstract

Post-Traumatic Stress Disorder (PTSD) PTSD can occur at any age and can follow a natural disaster such as flood or fire, a man-made disaster such as armed conflict such as those experienced in Somali, Rwanda, Burundi, Sudan, and many other parts of Africa. Other causes of PTSD may include imprisonment, assault, domestic abuse or rape of victims during conflict. The causes of PTSD are still under investigation. Many scholars have discussed and theorized extensively on the its causes, the National Institute of Mental Health points to psychological, genetic, physical, and social factors as its possible abode. The amygdala, a structure in the brain, is part of the limbic system that controls emotions such as fear, autonomic reactions (e.g., increased heart rate and blood pressure, the startle response), and memory. Dysfunction in this structure may produce symptoms of PTSD. The various physiological markers which have been studied in PTSD patients include heart rate, blood pressure, muscle activity (with electromyography), electrodermal activity (EDA), electroencephalography and peripheral temperatures. Heart rate and EDA are the most commonly studied parameters and point towards the involvement of the sympathetic nervous system. PTSD alters the way the human body responds to stress, effecting mediators such as stress hormones and neurotransmitters. Prior exposure to trauma may increase risk, suggesting a kind of learned response. Drumming as form of music therapy plays a protective role. The paper explores some case scenarios from both biomedical and phenomenological experiences.

Introduction

Post-Traumatic Stress Disorder (PTSD) can be an extremely debilitating condition that can occur after exposure to a terrifying event in which both physical and psychological harm either occurred or was merely threatened. Traumatic events that trigger PTSD include but are not limited to violent and armed ethnic conflicts like those in Somalia, Rwanda, Burundi, and Sudan. Others may include personal assault such as rape or mugging, natural or human-caused disasters, accidents, or military service in combat zones such as that experienced by the veterans of Vietnam, Korea, Iraq and Gulf Wars, or those who served as rescue workers during the aftermath of disasters such as the 9/11 attacks on the World Trade Center in the United States and the 1998 Nairobi/Tanzania US Embassy Bombings. Other factors that lead to PTSD may include automobile accidents, physical and sexual assault, and refugees fleeing violence in their home countries among others. Effective treatments have now been developed to help people cope with PTSD. Research is also helping more scientists to better understand the condition and how it affects both the human brain and body. Different forms of music such as drumming are becoming an important therapeutic tool. This paper examines the role of African drumming in the mitigation of PTSD.

The Use of African Drumming

Among the many African ethnic groups, regular and balanced meter are regarded as a sign of good health. Even in improvisations, the performers are expected to render an exact replica of a standardized musical practice. These musical practices that relay regular and replicated rhythms to heal the person in an immediate and powerful way by removing blockages and releasing tension can be seen in the performance of a Kenyan ethnic ritual dance, *ngoma* of the waTaita people. During this performance, a glissando is played by the lead drummer by gliding his left hand from the middle of the drum to the edge (*kusira ngoma*). By doing this, the drummer not only provides an expressively emotional pattern at the climax of the healing ritual but also provides a functional significance to the healing process because it is during this moment that the drummer sedates the *pepo* spirit to descend and exorcise the evil spirits from the patients.

Kusira ngoma, which literally translates into "going beyond with music," is the climax of the healing ritual and its ultimate extreme. This is the stage at which the patients shiver, fall to the ground and ultimately go into trance. During this healing ceremony, the master drummer controls the emotions of the patient while the patient unlocks his or her inner subconscious mind. In the middle of the performance, when the interlocking parts become intense, the patient is induced to a state where they start to dance pathogenically as they respond to the *mwazindika* drum, letting their souls soar into the supernatural world to meet the deity. In a similar supernatural meditation, Cornelius (1990, 127) found that the Afro-Cuban *bata* drums were believed to be capable of talking and communicating directly with the *orishas*, Yoruba gods. This power of the drum to be able to speak is also possibly seen as a catalyst to helping people to talk. Ruth Noonan, a practicing music therapist in Longmont United Hospital in Colorado, has observed that in her recent practice, she has witnessed the drumming help a patient to regain speech:

> There have been magical moments as well. One man who did not speak due to a stroke is now able to tell me that he grew up on a farm and always loved music. Now he is participating for the first time in music instead of just listening and he loves it... and he is talking (Noonan, 2005)

Among the waTaita people of Kenya, the approach to treatment of PTSD takes several forms, but one of the most significant treatment approaches for which this paper is devoted is the use of drumming. As the patients delve into subconscious worlds, the music healer plays *glissando* (*kusira ngoma*) on the *mwazindika*. The music healer, who occasionally wets his left thumb with saliva and glides it from the edge to the middle of the drum, continues to pound from the edge to the middle with his right hand until the patient stands and gets out of the healing arena.

Cross-Cultural Validity of Posttraumatic Stress Disorder

Cross-cultural studies of health practices and anthropologic studies of health culture can make significant contributions to the field of complementary and alternative medicine (CAM). Contemporary medical practice has become increasingly discordant with lay expectations. Modern physicians diagnose and treat diseases (abnormalities in the structure and function of body organs and systems), whereas patients suffer illnesses (experiences of disvalued changes in states of being and in social function; the human experience of sickness (Eisenberg, 1976; Enoel, 1977; Feinstein, 1967). Post-traumatic stress disorder is defined as a severe reaction to a traumatic event involving actual or threatened death or serious injury to self or others. According to the mental institute of health, PTSD can have an acute onset soon after the trauma, or a delayed onset in which the symptoms occur more than six months after the trauma. It has been observed that an acute PTSD may resolve after three months, followed by a chronic form of the disorder which may persist even past that time. It should be noted that PTSD can occur at any age and can follow a natural disaster such as flood or fire, or a man-made disaster such as armed conflict, such as those experienced in Somalia, Rwanda, Burundi, and Sudan. Other PTSD-linked traumatic events may include imprisonment, assault, domestic abuse or rape.

The causes of PTSD are not quite well-known. While many scholars (Konchin 2001; Shay 1995; Kessler, et al. 2008) have discussed and theorized extensively on its causes, the National Institute of Mental Health points to psychological, genetic, physical, and social factors as its possible abode. Literature on mental health has shown that the amygdala, a structure in the brain, is part of the limbic system that is involved in the expression of emotion – especially fear, autonomic reactions (e.g. increased heart rate and blood pressure, the startle response), and emotional memory. Dysfunction in this structure may produce symptoms of PTSD (Kessler, Galea, Gruber, Sampson, Ursano, et al., 2008). PTSD alters the way our body responds to stress, affecting mediators such as stress hormones and neurotransmitters. Prior exposure to trauma may increase risk, suggesting a kind of learned response. Music therapy plays a protective role. In studies of Vietnam veterans, drumming exercises greatly reduce stress among them and other victims of trauma by altering their brainwave patterns.[1]

The aim of treatment is to reduce the symptoms by encouraging the affected person to recall the event, to express feelings, and to gain some sense of mastery over the experience. The drum has been used in many communities, especially in Africa, to help the victim recall the traumatic event and dance away the spirit. As a catastrophe, famine is one seasonal event that brings trauma to families all over the world. Among the Luo ethnic group of Kenya, each famine is allocated a name. This name is also synonymous with the bad spirit that is responsible for the famine. One such famine is *nyawawa* which emerges as a wind and sweeps across the Luo country. This wind is a spirit and may cause death and destruction, including bringing famine. Whenever families think of famine, they remember their loved ones who may have died due to lack of adequate food because of this spirit. In many parts of the Luo ethnic group, the bad forces of nature that stop the rains and bring starvation are also responsible for causing other sicknesses as well. The National Institute of Mental Health identifies three stages of PTSD. The first stage is where the patient experiences recurrent distressing memories of the event or experiences recurrent dreams of the event, or flashback episodes when the event seems to be reoccurring. The Luo community will play drums in such seasons to exorcise the spirits from the affected children and adults alike.

This approach is consistent with the National Institute of Mental Health's Behavioral Therapy which is used to treat avoidance symptoms. Behavioral techniques used include the graded exposure to rhythms and melodies alike. Just like medications act upon the central nervous system to reduce feelings of anxiety and associated ailments, so the drums elicit that effect. Antidepressant medications have been proven effective in treating PTSD, including newer compound agents such as Prozac, Paxil and Zoloft. Sedating agents can help with sleep disturbance. Anti-anxiety agents may be

[1] Friedman, Dorian, *U.S. News & World Report*; 06/09/97, Vol. 122 Issue 22, p17, 1/2p, 1c

useful, but the benzodiazepines, a common class of these medications, can potentially become addictive due to dopamine release. Dopamine is a neuromodulator that is used by neurons in several brain regions involved in motivation and reinforcement, most importantly the nucleus accumbens (NAc). Dopamine can alter the sensitivity of its target neurons and other neurotransmitters such as glutamate, which can trigger motivational stimuli, which encourage a person to perform or repeat a behavior. The role of dopamine in repetitive behavior depends on three factors: (1) the type of stimuli that activate dopaminergic neurons, (2) the specific brain area(s) affected by dopamine, and (3) the mode of dopaminergic neurotransmission and whether it is phasic-synaptic or tonic-nonsynaptic (Di Chiara 1995). The Luo drum patterns have similar effects on the NAc. The second stage of PTSD (as defined by the NIMH) is the avoidance where the patient exhibits significant inability to recall important aspects of the trauma, or lack of interest in activities. The patient also shows feelings of detachment, sense of a foreshortened future, psychic or emotional "numbing," and restricted range of affect. The third and last stage is the arousal stage where the patient is highly irritable or shows outbursts of anger. The victim also exhibits sleeping difficulties and has difficulty concentrating leading to Attention Deficit Disorder (ADD), especially in children. Early counseling and crisis intervention through music are important for people who have experienced extremely stressful situations. Musical interventions may help prevent chronic forms of PTSD and should be part of public health responses to groups at risk, such as disaster victims.

Experience from a modern-day health facility

As part of my doctoral studies in the use of music in health with Dr. Miriam Zach of the University of Florida School of Music and Dr. John Graham-Pole, a pediatric oncologist at the University of Florida School of Medicine, I devoted two semesters to the Intensive Pediatric Care Unit (IPCU) of Shands Hospital of the University of Florida. I had been inspired by literature from The Harvard Medical School that highlighted drumming as having become an important therapeutic regimen for post-traumatic patients. During my work with the pediatric patients, I was able to play miniature "drum circles" and other forms of drum therapy to help children cope with their pain. In one instance, my drumming was instrumental in helping the nurse give routine injections to the children. Although my drumming temporarily reduced stress and fear associated with the jab and the piercing of the epidermis, this therapeutic intervention greatly helped the children. The effect was achieved by altering the children's brainwave patterns.

On several occasions, while I was at the IPCU, I had a dozen child residents with disorders of the nervous system, disorders of the brain, meninges, and skull gathered for their weekly rhythm circle. With the help of therapist Kathy Dewitt, Rusti Brandman and John Graham-Poe, I had each child shake maracas to the tunes of Sani Bona, a South African tune. Some who seemed otherwise confused could nonetheless tap perfectly on cue. Patients unable to speak were able to sing childhood songs. Those barely able to walk could lift their limbs when the dancers moved to my drum beats. As of yet, neuroscience has no sure explanation, but some experts think that the brain's receptors for music and rhythm are spared the early ravages of senility. And while many scholars may argue that no amount of drumming can cure the disease with significant results, drumming can improve the quality of life and offer another way of pain relief and attainment of the highest satisfaction.

Conclusion

The effect of drumming in the treatment of diseases should not be disputed. Since our ancestors first struck sticks and rocks against the ground, drumming has been a sacred ritual in many societies.[2] This belief emanates from the fact that throughout the world, the drum has been used for healing purposes. The traditional peoples of Africa, the Aboriginals of Australia, the Balinese of Southeast Asia, the Native American Indians, and the ancient Celts among others all used drumming to bring the rain, the sun, a bountiful harvest, successful hunting and

[2] Friedman, Dorian, *U.S. News & World Report*; 06/09/97, Vol. 122 Issue 22, p17, 1/2p, 1c

good health.[3] The drum has also been used in tribal societies with shamanistic traditions while communicating with the gods. In West African wisdom teachings, Cottel (2001) noted that an emotional disturbance manifests as an irregular rhythm that blocks the vital physical energy flow. Cottel also refers to current medical research which has shown that stress is a cause of ninety-eight percent of all diseases such as heart attacks, strokes, immune system breakdowns, among others. Recent biofeedback studies (for example, Spintge 1992; Harner 1990; McIntosh 1996) show that drumming along with our own heartbeats alters brainwave patterns (increasing alpha) and dramatically reduces stress. Unlike the Western cultures which rely on material evidence such as infection from bacteria or viruses, cell production such as cancer, or genetic anomalies such as defective chromosomes, the non-Western cultures relate to diseases from a cultural perspective connecting their etiology to the metaphysical world. These non-Western cultures' understanding of the etiology of disease is embedded in their cosmology. For example, the Luo tribesmen of Kenya believe that HIV/AIDS is caused by a curse. In this perspective, a curse is viewed as evil, pronounced or invoked by another living person, or the spirit of the dead. Among the Luo ethnic group, drum ensembles are performed with the object of exorcising the bad spirit from patients. This review shows that culture and etiology of disease are intricately intertwined, hence a multifaceted approach to the diagnosis is as essential to the prognosis.

[3] Moonsong Music Company at http://www.holistic.ie/main/essays/music.htm

References

Akombo, D. (2017). Effects of Music Practice on Anxiety in Undergraduate Music Students. In *Proceedings Journal of the Interdisciplinary Society for Quantitative Research in Music and Medicine*. Pg. 118-130.

anon (1994) Diagnostic and Statistical Manual of Mental Disorders. Vol. IV. American Psychiatric Association. 1994

anon (1998) Post-Traumatic Stress Disorder (PTSD).

Akombo, D. (2001) "Reporting on Music Therapy in Kenya". In Voices: A World Forum for Music Therapy, 1 (1), Oslo: Sogn og Fjordane University Press.

Akombo, D. (2017). Effects of Music Practice on Anxiety in Undergraduate Music Students. In *Proceedings Journal of the Interdisciplinary Society for Quantitative Research in Music and Medicine*. Pg. 118-130.

Bell, Taunjah & Akombo, David (2017). Long-term music-listening's effects on blood pressure, heart rate, anxiety, and depression. *Journal of Alternative Medicine Research, 9*(1) Available at https://www.novapublishers.com/catalog/product_info.php?products_id=60693.

Cornelius, S. (1990) "Encapsulating Power: Meaning and Taxonomy of Musical Instruments of Santeria in New York City". In Selected Reports in Ethnomusicology (8) 125-142.

Cornelius, S. (2000) "They Just Need Money: Goods and Gods, Power and Truth in a west African Village". In The African Diaspora, Ingrid Monson (ed). New York: Garland Publishing Inc.

Cottrell, A. (2001) Healing Musical Instruments.

Di Chiara, G. (1995). The role of dopamine in drug abuse viewed from the perspective of its role in motivation. Drug and Alcohol Dependence 38: 95-137.

Eisenberg L: (1976). Conceptual models of physical and mental disorders, in *Research and Medical Practice*. Amsterdam: Associated Scientific Publishers. pp. 3-23.

Enoel G.L (1977). The need for a new medical model: A challenge for biomedicine. *Science 196*:129-136.

Erdtsieck, J. (1997) Pepo as an Inner Healing Force: Practices of a Female Spiritual Healer in Tanzania. Amsterdam: Royal Tropical Institute.

Fedders, A. & C. Salvadori (1979) Peoples and Cultures of Kenya. Nairobi: Transafrica.

Feinstein, A.R (1967). *Clinical Judgment*. Baltimore, Williams and Wilkins. pp, 24-25.

Finke, J. (2001) Traditional Music and Cultures of Kenya.

Finke, J. (2002) Music in Taita. Personal Communication, January 6th, 2002.

Friedman, Dorian, *U.S. News & World Report*; 06/09/97, Vol. 122 Issue 22, p17, 1/2p, 1c.

Friedson, S. (1998) "Tumbuka Healing". In The Garland Encyclopedia of World Music Vol. 1, R.M. Stone (ed). New York: Garland Publishers.

Good, Charles M. (1987) Ethnomedical systems in Africa: patterns of traditional medicine in rural and urban Kenya. New York: Guilford Press.

Harper, P. (1969) "Dance In Nigeria". In Ethnomusicology, 9 (1), 280-295.

Hobart, A., U. Ramseyer & A. Leeman (1996) The Peoples of Bali. Oxford: Blackwell Publishers Limited.

Hollandsworth, J. (1990) The Physiology of Psychological Disorders: Schizophrenia, Depression, Anxiety, and Substance Abuse. New York: Plenum Press.

Hood, M. (1982) The Ethnomusicologist. Kent: Kent State University Press.

Horden, P. (2000) Music as Medicine: The History of Music Therapy since Antiquity. Hants: Ashgate Publishing House.

Huebner, P. (2001) Nature's Laws of Harmony in the Microcosm of Music. From a lecture given by the classical composer Peter Huebner at the medical faculty of the University of Heidelberg, the University of Tel Aviv and the University of Magdeburg.

Janzen, John M. (1992) Ngoma: Discourses of Healing in Central and Southern Africa. Berkeley: University of California Press.

Janzen, John M. (1978) "The Comparative Study of Medical System as Changing Social Systems". In Social Science and Medicine (12) 2B: 121-129.

Kessler, R., Galea, S., Gruber M., Sampson, N., Ursano, R., & Wessely, S. (2008). Trends in mental illness and suicidality after Hurricane Katrina. *Mol Psychiatry 13*(4):374-84.

Konchin, D. (2001) Post Traumatic Stress Disorder: the invisible injury. Oxford: Success Unlimited Publishers.

Kovach, A. M. (1985) "Shamanism and Guided Imagery and Music: A Comparison". In Journal of Music Therapy, XXII (3). National Association of Music Therapy, Inc.

Moreno, J. (1995) "Ethnomusic Therapy: An Interdisciplinary Approach to Music and Healing." In The Arts in Psychotherapy 22(4), 329-338.

Mullings, L. (1984) Therapy, Ideology, and Social Change: Mental Healing in Urban Ghana. Berkeley: University of California Press.

Nketia, K. (1962) "The problem of Meaning in African Music". In Ethnomusicology. Ann Arbor, Society for Ethnomusicology.

Nketia, K. (1962) African Music in Ghana: A survey of Traditional Forms. Accra: William Clowes & Sons Limited.

Nketia, K. (1979) The Music of Africa. London: Victor Gollancz Ltd.

Noonan, Ruth (2005). Personal Interview, Music therapist in Longmont United Hospital in Colorado.

PTSD Pharmacotherapy (2007). VA/DoD Clinical Practice Guidelines. http://www.oqp.med.va.gov/cpg/PTSD/ G/Interv_Sum508.pdf. Accessed on June 8.

Robert, S. & M. Louise (1961) Songs of power. Syracuse: Church of the Christian and Missionary Alliance.

Senoga-Zake, G. (1986) Folk Music of Kenya. Nairobi: Uzima Press.

Shay, J. (1995) Achilles in Vietnam: Combat Trauma and the Undoing of Character. New York: Touchstone/Simon & Schuster.

Re-Examining Standard Tuning and its Effects on Humans

L. Elise Carter

Abstract

Is one tuning better than another when it comes to tuning an instrument? The Schiller Society promotes tuning instruments to 'A' = 432 Hz. Another author, Leonard Horowitz, claims that 'A' = 444 Hz is a preferred tuning. To find out if there is scientific evidence to these claims, the author did a full brain map on herself to look at changes in brainwaves during these different conditions. She looked at Delta, Beta, Alpha and Theta waves while listening to a single tone of C. This tone was played on the flute and tuned to each of three tunings – the two above-mentioned tunings, A = 444 Hz and A = 432 Hz, plus A = 440 Hz. A baseline reading was taken and was compared to each of the three tunings. All four brainwaves were examined and the locations where these brainwaves were most concentrated. The author found that there was a rise in the amount of alpha waves upon listening to the tuning of 'A' = 440 Hz and 'A' = 444 Hz in the frontal lobe of her brain.

Introduction

Most musicians take out their tuner and tune to A = 440 Hz without thinking. However, this standard of tuning was not universally used until after World War II. During the year 1859, for the first time, a standard of tuning was set – for A = 435 Hz in Paris. "At a time when the tuning fork was vibrating to the 435 Hz frequency in Paris by ministerial decision; and to 439 Hz in London by royal decree, the renowned Italian composer Giuseppe Verdi adopted himself the 432 Hz frequency" (Comte, E., 2013. para. 7). Before then, there were various tunings ranging from 417 to 460 Hz. Some composers had their preferred tunings; Mozart liked A = 422 Hz, while Haydn preferred A = 423 Hz (Comte, 2013). A tuning fork used by the manufacturer of Mozart's pianofortes was tested to be A = 421.6 Hz (Gioia, 2008).

Does the frequency that we are tuning to matter? Many people think it does, and many of those people do not like A = 440 Hz. There is a whole society devoted to reviving Verdi's tuning, called the Schiller Society (The Schiller Institute, 2017). There is another movement initiated by a man named Leonard Horowitz who feels that A = 444 Hz is a sacred tuning, and that it promotes miracles and love (Horowitz & Puleo, 2001). At this tuning, when one plays the note C, it registers at 528 Hz. This exact Hertz frequency is the pitch that Horowitz feels is a tone of miracles and love, and his basis for this came from decoding the Bible (Horowitz & Puleo, 2001). Horowitz has gone so far as to make his own radio station that will play music that has been altered from the tuning at A = 440 Hz to A = 444 Hz. The author decided to measure her brain response to hearing C at 528 Hz, and compare it to hearing C at the Verdi tuning and a C at the standard 440 Hz tuning. The author found an increase in alpha waves with both the 440 Hz tuning and the 444 Hz tuning.

The first brainwaves that were discovered were the alpha waves in 1924 by Hans Berger (Empson, 1986). He noticed that when the eyes are closed these waves at 10 Hz were more prominent than with the eyes open. Since then, alpha waves have been one of the most-studied brainwaves, and there are numerous ways of describing them. Foster described alpha waves as "a state of relaxed wakefulness in which attention tends to wander, free association is enhanced, and behavioral efficiency of routine reactions and creative thought is good" (Foster, 1990). Alpha waves are present with light meditation with the eyes closed (Miller, 2011).

Neurofeedback is the practice of using the EEG readings to alter brainwave production so that a desired state of being can be achieved. The first neurofeedback began when two different researchers, Kamiya and Stern, both realized that an animal or person can change the brainwaves they are producing when given a reward. (Miller, 2011, Foster, 1990) There are numerous neurofeedback practitioners now using neurofeedback to treat depression, ADHD, alcoholism, OCD, and Peak performance training (Demos, 2005).

According to John Demos, alpha waves are something that should be uptrained or downtrained depending on the client. Most often clients with anxiety have not enough alpha or an imbalance of alpha from the right to the left hemisphere. Too much alpha may cause insomnia. An imbalance of too much alpha on the left hemisphere in relation to the amount of alpha on the right hemisphere can be a symptom of depression, and more alpha at the frontal region of the brain is a preferred state to having high alpha in the posterior region of the brain (Demos, 2005). Alpha waves also have an increased presence in the brains of those who meditate a lot, as well as athletes and musicians (Bergland, 2015).

Method

Using the 19 channel EEG Neurofeedback software, the author did a reading of her brainwaves during four conditions; baseline, and then three recorded pitches of the note C. These were pitched using the A = 440 Hz tuning, the A = 444 Hz tuning and the A = 432 Hz tuning. The subject's brainwaves for each condition were recorded with a cap of 19 electrodes placed on the international 10-20 locations of the brain. An assistant activated an iPad with the recorded tones on it so that the subject would not know which tone was played for each condition. A chart of the actual hertz for the three C's is shown in appendix A. The author examined the four basic brainwaves: Delta, Theta, Alpha and Beta at each location of the 10-20 electrodes, and examined the data using the NeuroGuide software and the NeuroGuide normative database.

Results

Alpha Waves

The most striking changes between the baseline condition and the other three conditions occurred with the alpha waves. The subject's baseline reading of alpha waves was higher than average but not significant, with a standard deviation of 1.5-2.0 (shown in a light orange-yellow color on the brain map – fig. 1 – Appendix A). When condition 440 = Hz and condition 444 Hz were sounded, it resulted in alpha waves that registered between 2.0 and 3 deviations from the normative database (shown in a dark orange color on the brain map). The 10-20 poles that registered this change in alpha waves were FPz, FP1, FP2, F3, F4, T3, C3, Cz, C4 and T4. These changes were seen in the 440 Hz condition and the 444 Hz condition, not the 432 Hz condition or baseline. These changes are shown in the Relative Power Z Scores brainmap (Fig. 1).

APPENDIX A

Tuning Chart (Suits, 2015)

A = 440 Hz	B = 493.88 Hz	C = 523.25 Hz
A = 444 Hz	B = 498.37 Hz	C = 528.01 Hz
A = 432 Hz	B = 484.90 Hz	C = 513.74 Hz

Figure 1

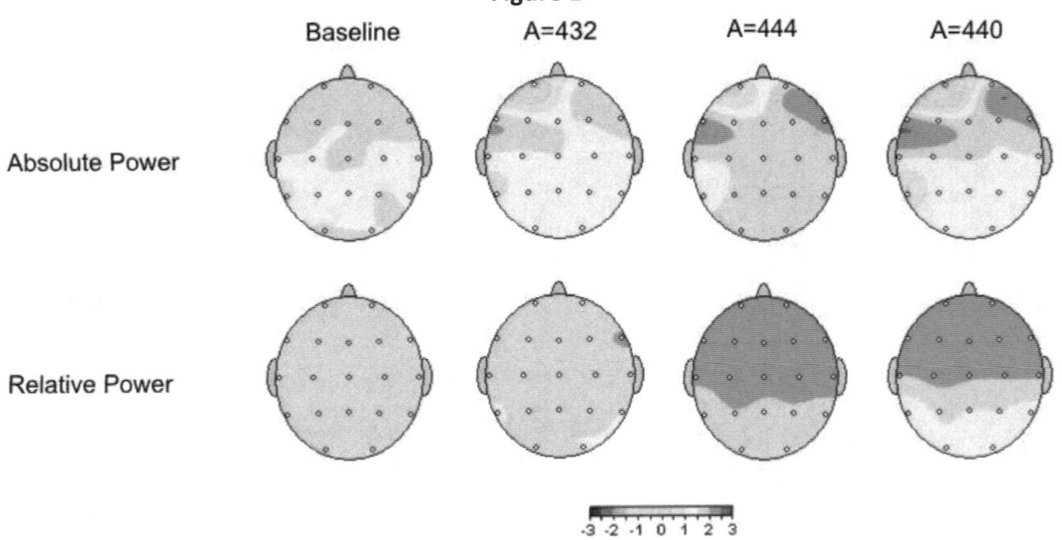

In examining the relative power Z scores at each 10-20 location on the brain, *every* location on the brain showed higher amounts of alpha waves with condition 444 Hz compared to baseline condition. When compared to 440 Hz, the 444 Hz condition had higher amounts of alpha at every pole except Fp2, F4 and Fz. (Figure 2).

Figure 2. Relative Power Z Scores

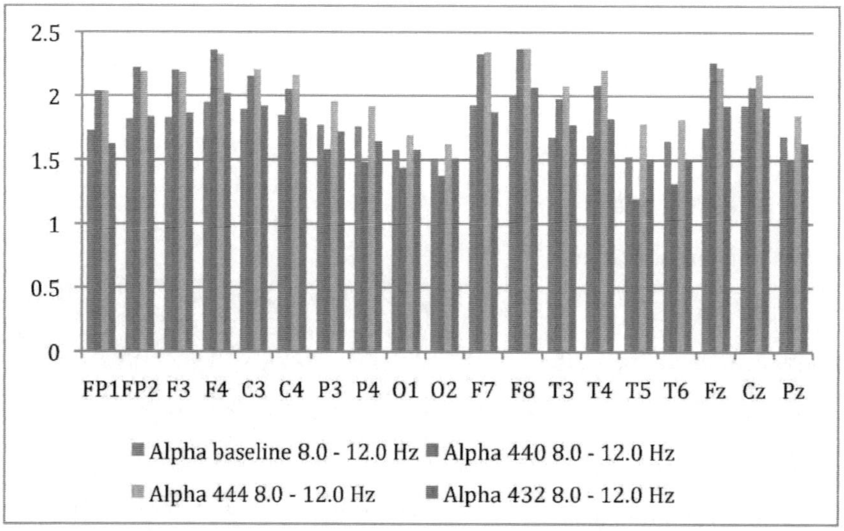

Figure 3. Absolute Power Alpha Z Scores

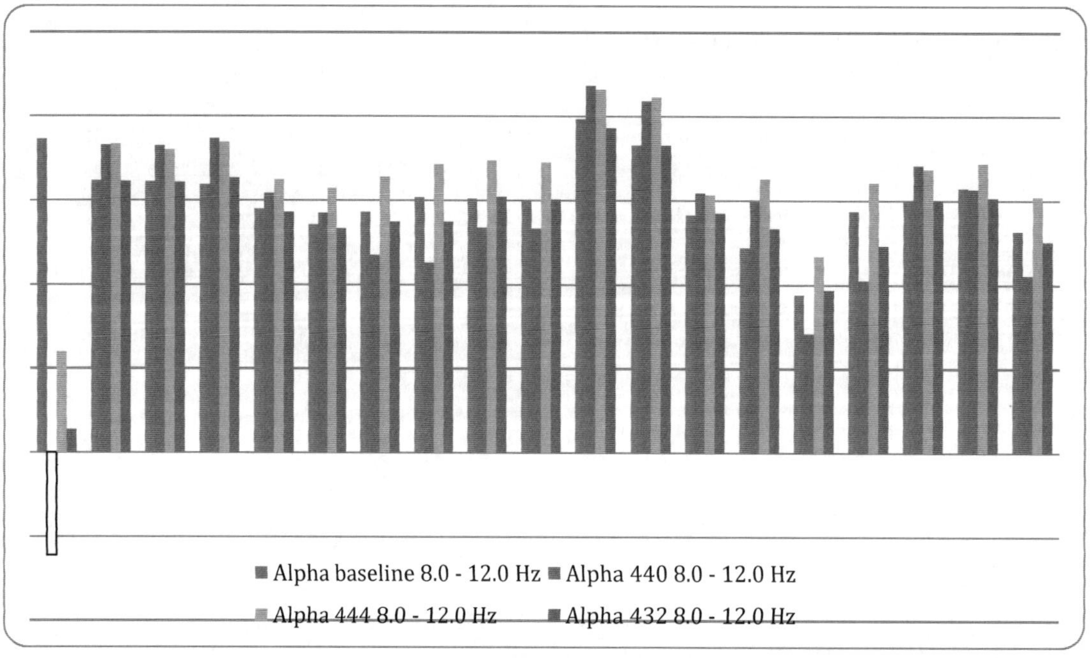

In examining the Absolute power Z scores, the areas that showed the change in alpha waves were located on and around the F7 and F8 poles. (See fig 1 and fig 3.)

A two-tailed t-test was done for each condition at pole F8. In order to be significant, p must be less than 0.05. The results were as follows: baseline condition p = 0.067325, for condition 440 Hz, p = 0.036501; for condition 444 Hz, p = 0.034362; and for condition 432 Hz, p = 0.0677. Therefore, the conditions of 440 Hz and 444 Hz were significantly higher compared to the normative database.

F7 location p values were as follows: Baseline: p = 0.04711; 440 Hz: p = 0.029036; 444 Hz: p = 0.03092; and for condition 432 Hz: p = 0.053508. Condition 432 Hz was not significantly different from the normative database, but all the other conditions were.

Beta waves

The subject had beta waves between 1.5 and 3 standard deviations below the normative database at every pole. This remained constant throughout all conditions.

FP1 Pole

For Theta, Delta and Beta 12. 00 – 25.00 Hz, the FP1 pole showed significantly lower than normal waves than the normative population at the three tone conditions of 440 Hz, 444 Hz, and 432 Hz. Baseline condition was near 0. In the absolute value alpha chart, condition 440 Hz was below the normative population even though the three other conditions were above the normative population. The reading at the Fp1 pole is frequently disturbed by eye or forehead movement, called movement artifact. (Demos, 2005) Because this pole showed such unusual readings throughout all of the conditions, the author feels that the unusual readings were due to artifact.

Theta Waves

Nothing was significantly different from the normative database in absolute power among all four conditions.

Delta Waves

Nothing was significantly different from the normative database.

Discussion

When the tones at 444 Hz and 440 Hz were sounded, the subject reacted with an increased amount of alpha waves in the frontal lobe. These reactions were shown in the relative power Z scores. In EEG Neurofeedback, often the relative brain maps are the scores that are used for neurofeedback training (Demos, 2005). The frontal lobe is considered to be the social brain, this has locations that are connected with speech, emotional processing, regulation, and expression of emotion, such as inhibition or lack of inhibition. The frontal lobe also has connections to the amygdala. Therefore, the conditions A = 444 Hz in the frontal lobe of the subject increased the amount of alpha waves present at every pole compared to the baseline condition.

If listening to a single tone can produce a state of increased alpha waves in the listener, then Leonard Horowitz's theory that C = 528 Hz is a special tone may have some truth behind it. Both the tones A = 440 Hz and A = 444 Hz produced an increase in alpha waves in the frontal region of the brain, but the A = 444 Hz tone produced more alpha in every 10-20 location except the three previously mentioned.

Alpha waves are also shown to stimulate creativity and reduce depression. If the brainwaves of the subject were increased just by listening to a single pitch, then perhaps this pitch is special indeed. Further research is needed to determine if alpha waves would be produced among non-musicians listening to this pitch. Further research is needed to determine whether the specific areas of F8 and F7 would be activated in other people.

In the absolute power brain map, the areas of F8 and F7 had higher amounts of alpha. The area of F8 is associated with Brodman area 47. Brodman area 47 is an area for language, but also music processing, auditory processing, emotional processing and sustained attention. The pitches that the subject listened to for the experiment were pitches that she had recorded on the flute herself. When the assistant played the recorded pitches, during the experiment, the subject thought she knew which pitch was sounded as she listened to them. She recognized these pitches, not necessarily by the sound, but by the way she felt while listening to them. F8 could be the area that her brain was using in order for her to identify the pitches, and that was why this had the absolute highest output of alpha waves. She used her emotional reaction rather than auditory sounds to identify which tone was played. In addition to this, she had been working with the different pitch centers because of her curiosity about 'A' = 444 Hz for the last few years.

In particular, the area of F7 is associated with the location of Broca's area. This is an area of the brain that is used in speech production. Motor planning for swallowing and the movements for speech production come from this area. (McCaffrey, 2014) The author played the flute when she recorded the tones. When changing pitch, a flutist drops the position of the jaw to make the pitch lower. When playing the flute, the position of the tongue in the mouth will also produce different tone colors as well as raising or lowering the pitch of the instrument. Perhaps the areas in the brain used to activate the motor planning necessary for pitch changes were activated when she listened to the flute pitches. Perhaps these areas were not activated at A = 432 Hz because the author does not perform regularly at that pitch level. Further research is needed to determine whether this increase in alpha at the F7 and F8 poles is something that would be seen in the general population.

As a musician and someone who also meditates frequently, the author's brain showed a high amount of overall alpha. She had a very strong reaction to hearing the conditions of 440 Hz and 444 Hz. Hearing these tones caused much more alpha than the pitch of 432 Hz. The rise in alpha would indicate that these pitches are important. Further research is needed to determine if the particular note C was the reason for increased alpha or whether it was the tuning to 'A' = 440 Hz and 'A' = 444 Hz. If another pitch were played at the same tuning of 'A' = 440 Hz and 'A' = 444 Hz, would the subject's brain still react with more alpha waves? Would the subject's same areas be activated if she was listening to that pitch produced on another instrument or from a computer-generated sound?

Limitations

The author was the only subject in the study; therefore, it is hard to know whether the results were specific to her, or were something that the general population would experience. The author suspects that the readings at the Fp1 pole were due to artifact at the Fp1 pole site, so it is hard to know what the accurate readings might have been for this pole. The author has been working with the 440 Hz and 444 Hz conditions for the past several years, so she was more familiar with these tones than the average person or musician would be.

Implications for Music Therapy

If the author's brain is representative of the population, then playing music in the key centered around the pitch of C at the tunings of both A = 440 Hz and A = 444 Hz would cause a rise in alpha waves in the brains of the listeners. Increase in frontal alpha waves can be an EEG neurofeedback treatment for depression and anxiety. It could create a calm and meditative state in the listener. For a music therapy client who is agitated, anxious, or upset, music in the key of C major could create a more relaxed state in the client. Other keys centered around the note C could be used as well. Using the tuning of A = 444 Hz could increase the amount of alpha overall in the brain compared to A = 440 Hz. However, if a client has insomnia or an imbalance of alpha waves, using a tuning of A = 444 Hz or the key of C would not be helpful for them, and the therapist would have to be aware through verbal feedback and checking in with the client to find out their reactions.

More research would be needed to determine if just one note of C was the cause of the increased alpha, or if music centered around the key of C also increases the alpha waves. More research is needed to be done to determine if the brains of non-musicians would react in the same way as the brain of a musician.

References

Bell, K. (2017). EEG to Brainmap. Retrieved from https://www.youtube.com/watch?v=Qkp9FblYOTs

Bergland, C. (2015). Alpha Brain waves boost creativity and reduce depression. Psychology Today retrieved from https://www.psychologytoday.com/blog/the-athletes-way/201504/alpha-brain-waves-boost-creativity-and-reduce-depression

Comte, E. (2013). Is 'A' 432 Hz a myth? Retrieved from http://medson.net/A432hz-myth.html

De Boeck, C. (2010). Brain Functions. Retrieved from http://www.staalhemel.com/2010/brain-functions.

Demos, J. (2005). Getting Started with Neurofeedback. New York: W.W. Norton & Company.

Empson, J. (1986). *Human Brainwaves, the Psychological significance of the Electroencephalogram*. New York: Stockton Press.

Foster, D. (1990, 2003). EEG and Subjective Correlates of Alpha-Frequency Binaural-Beat Stimulation Combined with Alpha Biofeedback. Retrieved from http://www.stealthskater.com/Documents/Lucid_07.pdf

Gioia, T. (2015). Are we all mistuning our instruments, and can we blame the Nazis? *The Daily Beast.* Retrieved from http://search.proquest.com/docview/1695004832?accountid=12536

Horowitz, L., Puleo, J. (2001). *Healing Codes for the Biological Apocalypse.* Sandpoint, Idaho: Tetrahedron Publishing.

Jasper, H. H. (1958). The Ten-Twenty Electrode System of the International Federation. *Electroencephalography and Clinical Neurophysiology, 10 (1958),* pp. 371–375

Kamiya, J. (1969). Operant control of the EEG alpha rhythm and some of its reported effects on consciousness. In C. T. Tart (Ed.), *Altered states of consciousness* (pp. 519-529). Garden City, N. Y.: Anchor Books.

McCaffrey, Patrick (2014). Neuroanatomy of speech swallowing and language. Retrieved from https://www.csuchico.edu/~pmccaffrey/syllabi/CMSD%20320/362unit4.html

Miller, E. (2011). *Bio-Guided Music Therapy A Practitioner's Guide to the Clinical Integration of Music and Biofeedback.* London and Philadelphia, PA: Jessica Kingsley Publishers.

Schmidt, L. and Trainor, L. (2001). Frontal brain electrical activity (EEG) distinguishes valence and intensity of musical emotions. *Cognition and Emotion, 2001, 15 (4),* 487–500.10-20 sites and cortical functions. Retrieved from https://brain-trainer.com/answers/about-the-brain/10-20-sites-cortical-functions/

Suits, B.H. (2015). Physics of Music Notes. Retrieved from: https://pages.mtu.edu/~suits/notefreqs.html

Thatcher, R. (2016). QEEG Normative databases. Retrieved from: ANIPUBLISHING.com

Thatcher, R.W., Walker, R.A. and Guidice, S. Human. (1987). Cerebral hemispheres develop at different rates and ages. *Science, 236*: 1110-1113.

The Schiller Institute. (2017). The Science of Music. What is Beauty and Why is it Necessary? Retrieved from: http://www.schillerinstitute.org/music/main.html

Travis, F., Harung, H., Lagrosen, Y. (2011). Moral development, executive functioning, peak experiences and brain patterns in professional and amateur classical musicians: Interpreted in light of a Unified Theory of Performance. *Consciousness and Cognition. 20 (4)* DOI 10.1016/j.concog.2011.03.020

The Affective Cerebral Response to Music: Impact of Happy, Sad, and Angry Music

Chyi June Khoo

Abstract

What is happening to the brain when music triggers emotions? This single-subject pilot study examined the impact of different types of sounds on electroencephalogram (EEG) brainwave patterns. The researcher is interested in the influence of different kinds of music associated with emotions, specifically in the form of happy, sad, and angry music on electrical activity of the brain. The subject listened to nature sounds (as a neutral control condition) followed by a sequence of three musical selections that may be perceived as happy, sad, or angry. EEG was used to study the hemispheric lateralization pattern of these sounds. EEG analyses revealed responses during different affective stimulations on distinct brain regions of the cerebral cortex. Theta Absolute Power increased significantly at sixteen electrode sites while listening to happy music, and decreased significantly at ten electrode locations in sad music. During happy music, Theta/Beta ratio increased significantly at nine sites compared to neutral condition. Conversely, during sad music, Theta/Beta ratio decreased significantly at eight sites compared to control condition. Happy music, in this pilot study, seems to indicate the greatest increase in relaxation compared to neutral, angry and sad music, which may support image-based processing and the sparking of creativity. On the other hand, sad music may improve language-based processing and enhance intellectual activity.

Introduction

Music is a universal companion to people's everyday lives. Many of us use music to improve our health and wellbeing, such as to enhance relaxation, to help emotional regulation, and to release stress, among others. Individuals experience various bodily changes which include altered breathing, heartrate, galvanic skin response, and skin temperature in response to music (Landreth & Landreth, 1974; McFarland, 1985; Miller, 2011; Nakamura, 1984; Zimny & Weidenfeller, 1963). In addition to physical arousal, cortical arousal to music is also an area that has shown contribution to our health. Arousal in response to music correlates with emotional processing (Goldstein, 1980; Mitterschiffthaler et al., 2007; Panksepp, 1995; Sloboda, 1991). Emotions that music listeners tend to associate while listening to music are basic emotions such as happiness, sadness, anger, and fear (Schmidt & Trainor, 2001).

With the increase in development of, and interest in EEG and functional imaging in the past few decades, many studies have focused on brain activity to music. These studies have shown specific lateralization patterns to music arousal. EEG has indicated differences in hemispheric lateralization of emotions. Positive affect is found on left hemisphere activation while negative affect is found in right hemisphere activation (Davidson, 2002, 2004; Davidson et al., 1990; Tomarken et al., 1992). Moreover, the activity level of anger may be related to either hemisphere (Propper et al., 2010; Propper et al., 2011). Findings of brain lateralization in response to emotions in music are well represented. However, there is a lack of studies in brainwave patterns associated to hemispheric lateralization.

Theta wave activity (4-8 Hz) is associated with creativity, mystical experiences, inattention or spacing-out (in the case of ADHD children) (Miller, 2011). About one-third of normal young adults show <15 μV Theta wave activity in the frontal head regions which can be achieved by experiencing certain emotions, focused concentration, and mental tasks (Tatum, 2008). We can enhance the Theta wave activity by closing the eyes, elongating respiration and sleeping (Miller, 2011; Tatum, 2008). Since music-listening can evoke emotions and the experience will influence the Theta wave activity, which emotion among basic emotions will have effect on increasing the Theta wave activity? On the other hand, the Theta/Beta ratio may be utilized to gauge the relationship of meditative state to cognitive activity and for investigating attention-deficit/hyperactivity disorder (ADHD) (Miller, 2011; Monastra et al., 1999). In examining the Theta/Beta ratio, we may be able to gauge cognitive activity in relation to the meditative state. This research studies the Electroencephalogram (EEG) brainwave patterns observed in subjects listening to music associated with emotions, specifically in the form of happy, sad, and angry music. This is carried out by examining the subject's Theta wave activity and Theta/Beta ratio during the music-listening experience.

Methods

A single subject listened to nature sounds as a neutral control condition followed by a sequence of three musical selections that may be perceived as happy, sad, or angry, with the eyes closed. These musical selections were pre-rated in previous studies. A 19-channel EEG was recorded during these four musical excerpts for three minutes each with electrodes positioned according to the International 10-20 system (Figure 1). Neuroguide software was utilized to analyze the data. All records were visually scanned for artifacts. Trails with artifacts were removed from further analysis. NeuroStat (NS) was used to run statistical comparisons between the four conditions.

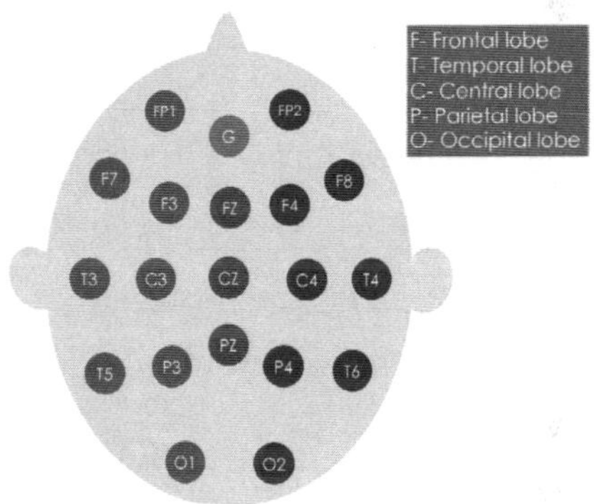

Figure 1: Shows the Electrodes Placement of International 10-20 System.

Results

Figure 2 shows the Z score of Theta Absolute Power and Relative Power compared to the normative database. Absolute Power is the measurement of energy output at each site and Relative Power is the distribution of energy resources being utilized. Zero on the scale is the mean. Warm colors are activities above the mean and cool colors are activities below the mean. When compared, the Absolute Power of the Sad condition is cooler than other conditions while the Happy condition is warmer than all the other conditions. On the other hand, the Absolute Power of the Neutral and Angry conditions are similar to each other. In terms of Relative Power, the distribution of energy in Neutral, Sad, and Angry conditions are a little under-utilized in the frontal and central lobe, relative to the Absolute Power.

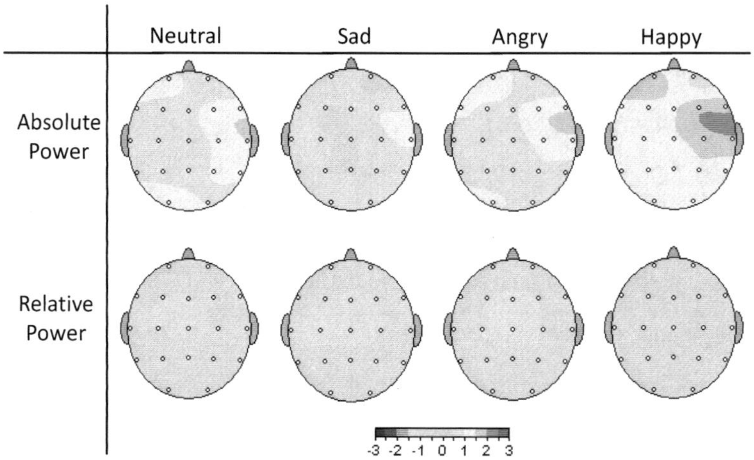

Figure 2: Shows the Theta Z scored FFT of Four Conditions.

Chart 1 shows the Theta Fast Fourier Transform (FFT) Absolute Power of four conditions on all electrode locations. EEG analyses revealed Theta FFT Absolute Power during Happy music increased

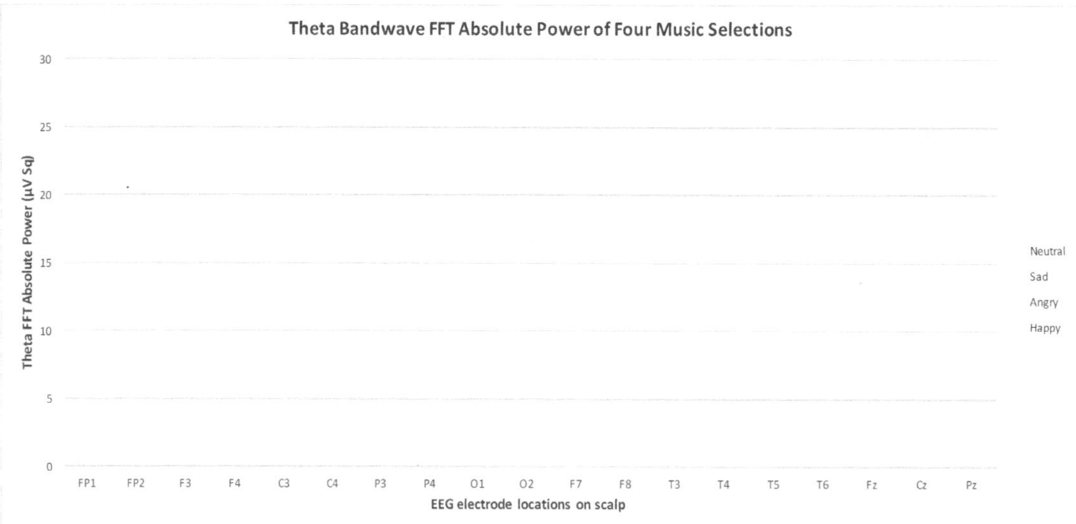

Chart 1: Shows the Theta FFT Absolute Power of Neutral, Sad, Angry and Happy conditions.

significantly ($p<0.05$) at all sites except for O1, O2, and T6, when compared to Neutral music. Conversely, during Sad music, Theta FFT Absolute Power decreased significantly ($p<0.05$) at FP1, FP2, F3, F4, C4, F8, T4, T6, Fz, and Cz when compared to control conditions. During Angry music, Theta FFT Absolute Power are similar to Neutral condition with P3, P4, O1, O2, T4, T6, and Pz somewhat lower.

Figure 3 shows the Theta/Beta ratio paired t-test of Happy, Sad and Angry conditions compared to Neutral condition. The grid shows the p-value of each comparison. Values below 0.05 are considered as significantly different. When compared, the Theta/Beta ratio of the Angry condition and the Neutral condition display no significant difference.

When the Happy condition is compared to the Neutral condition, their Theta/Beta ratio is significantly different towards the left frontal lobe of the brain (Figure 3). During Happy music, Theta/Beta ratio is significantly different with Theta increased at FP1 ($p=0.019$), F3 ($p=0.005$), F4 ($p=0.043$), C3 ($p=0.006$), C4 (0.009), F7 ($p=0.008$), F8 ($p=0.001$), Fz ($p=0.035$), and Cz ($p=0.011$) showing image-based processing towards the front of the brain. The Theta wave activity of the Happy condition increasing in the Theta/Beta ratio indicates the subconscious mind dominating thought processing. This also implies a more relaxed brain state during the Happy condition.

Comparing the Sad condition and the Neutral condition, the Theta/Beta ratio is significantly different towards the right side of the brain (Figure 3). During Sad music, electrode sites C4 ($p=0.013$), P4 ($p=0.021$), O2 ($p=0.035$), T3 ($p=0.016$), T4 ($p=0.008$), T6 ($p=0.011$), Cz ($p=0.005$), and Pz ($p=0.035$) are significantly different in Theta/Beta ratio. Beta increasing within the Theta/Beta ratio during the Sad condition

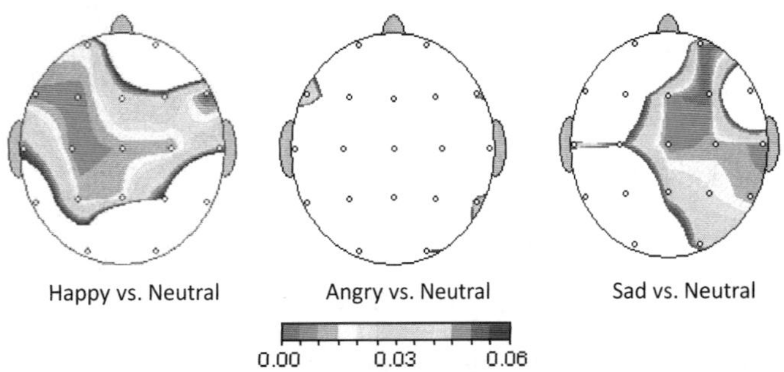

Figure 3: Shows the result of the paired t-test of Theta/Beta Ratio between Neutral and all the other music emotions.

suggests the conscious mind is taking over the subconscious mind. This is also an indicative of arousal (Miller, 2011).

Discussion

This study supports the previous studies of hemispheric lateralization in pleasant/happy music on the left hemisphere activation and unpleasant/sad music on the right hemisphere activation. In this pilot study, Happy music seems to demonstrate increased relaxation the most when compared to Neutral, Angry, and Sad music. Meanwhile, Sad music suggests an increase in alertness. This could be a significant factor for practitioners utilizing music selections in the therapeutic process. Music associated with different emotions has shown different neurologic responses. Music that elicits emotions is likely to impact the level of consciousness. An example will be Guided Imagery and Music (GIM) where the practitioner uses classical music as an agent into intersubjective interaction (Bruscia et al., 2005). While making music selections, considering the emotional aspects and the neurologic response to music may have important clinical impacts. Music therapists may look into using selections of Happy music to enhance image-based processing, spark creativity and support relaxation interventions. On the contrary, the use of Sad music that is related to cognitive activity may improve language-based processing, sustain concentration, and enhance intellectual activity during music therapy sessions.

In this study, Theta brainwaves and Theta/Beta ratios of Angry music were surprisingly similar to Neutral music. Previous studies have shown different findings about the emotion of anger being associated with hemispheric activities. One study shows that the emotion of anger relates to elevations in left anterior cerebral activation with approach tendencies (Carver and Harmon-Jones, 2009). Another study reported association of anger with relative right frontal cortical activity (Zinner et al., 2008). It has been suggested that anger is not lateralized within a hemisphere but related to how the hemispheres interact (Propper et al., 2011). It is likely that in this pilot study the interaction between two hemispheres are neutralized during the Angry and Neutral conditions. Deeper research into hemispheric asymmetry and

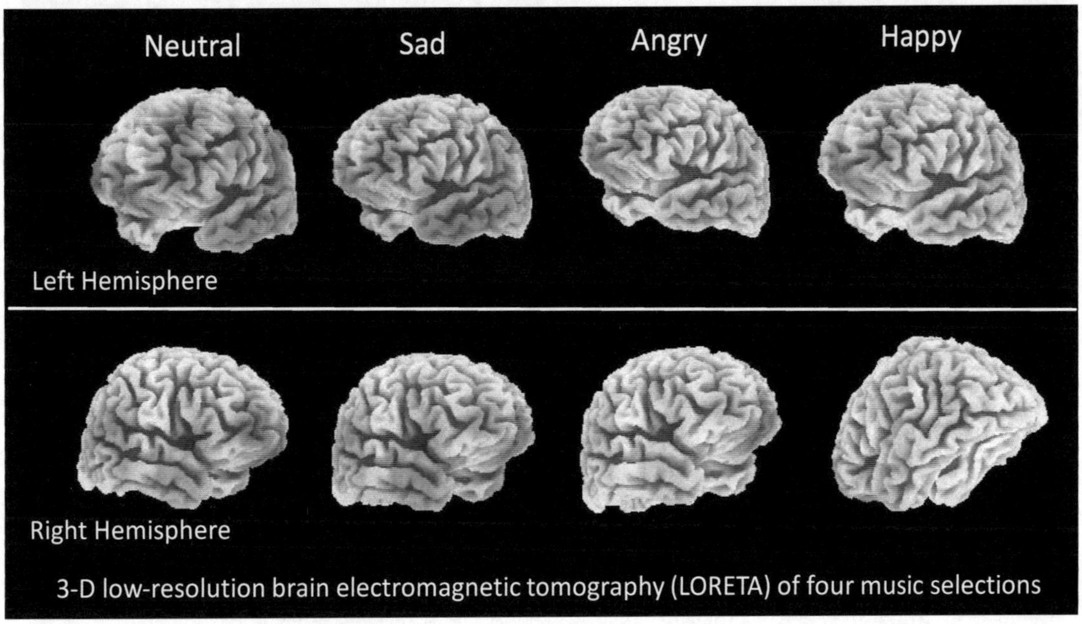

Figure 4: Shows the left and right hemispheres LORETA image of the four conditions.

coherence may be able to draw out further conclusions.

During this study, the researcher raised a question regarding the perception of emotion in music-listening. Is emotional processing in response to music different for everyone? Previous studies have indicated that there are individual differences in emotional reactivity, which are identifiable by features of affective style (Davidson, 1998). For future studies, self-reported affective state may provide a better understanding of the cognitive emotional processing of each individual. It has been suggested that the increase in hemispheric activity affects the increase in the propensity to experience a certain hemisphere's affective state (Propper & Brunyé, 2013). Future studies should consider self-reported subjective data in conjunction with objective data where their relationship can be examined.

Acknowledgements

The author would like to acknowledge Dr. Ruth E. Propper for paving the initial idea of this study and Dr. Eric B. Miller for assistance with this research project.

References

Bruscia, K., Abbott, E., Cadesky, N., Condron, D., Hunt, A. M., Miller, D., & Thomae, L. (2005). A Collaborative Heuristic Analysis of Imagery-M: A Classical Music Program Used in The Bonny Method of Guided Imagery and Music (BMGIM). *Qualitative Inquiries in Music Therapy*, *2*1-35.

Carver, C. S., & Harmon-Jones, E. (2009). Anger is an approach-related affect: evidence and implications. *Psychological Bulletin*, *135*(2), 183.

Davidson, R. J. (1998). Affective style and affective disorders: Perspectives from affective neuroscience. *Cognition & Emotion*, *12*(3), 307-330.

Davidson, R. J. (2002). Anxiety and affective style: role of prefrontal cortex and amygdala. *Biological Psychiatry*, *51*(1), 68-80.

Davidson, R. J. (2004). Well-being and affective style: neural substrates and biobehavioural correlates. *Philosophical Transactions of the Royal Society B: Biological Sciences*, *359*(1449), 1395.

Davidson, R. J., Ekman, P., Saron, C. D., Senulis, J. A., & Friesen, W. V. (1990). Approach-withdrawal and cerebral asymmetry: Emotional expression and brain physiology: I. *Journal of Personality and Social Psychology*, *58*(2), 330.

Goldstein, A. (1980). Thrills in response to music and other stimuli. *Physiological Psychology*, *8*(1), 126-129.

Landreth, J. E., & Landreth, H. F. (1974). Effects of music on physiological response. *Journal of Research in Music Education*, *22*(1), 4-12.

McFarland, R. A. (1985). Relationship of skin temperature changes to the emotions accompanying music. *Applied Psychophysiology and Biofeedback*, *10*(3), 255-267.

Miller, E. B. (2011). Bio-Guided Music Therapy: A practitioner's guide to the clinical integration of music and biofeedback. Jessica Kingsley Publishers.

Mitterschiffthaler, M. T., Fu, C. H., Dalton, J. A., Andrew, C. M., & Williams, S. C. (2007). A functional MRI study of happy and sad affective states induced by classical music. Human brain mapping, 28(11), 1150-1162.

Monastra, V. J., Lubar, J. F., Linden, M., VanDeusen, P., Green, G., Wing, W., & Fenger, T. N. (1999). Assessing attention deficit hyperactivity disorder via quantitative electroencephalography: an initial validation study. *Neuropsychology*, 13(3), 424.

Nakamura, H. (1984). Effects of musical emotionality upon GSR and respiration rate: The relationship between verbal reports and physiological responses. *Japanese Journal of Psychology*.

Panksepp, J. (1995). The emotional sources of "chills" induced by music. *Music*

Perception: An Interdisciplinary Journal, 13(2), 171-207.

Propper, R. E., Brunyé, T. T., Christman, S. D., & Bologna, J. (2010). Negative emotional valence is associated with non-right-handedness and increased imbalance of hemispheric activation as measured by tympanic membrane temperature. *The Journal of Nervous and Mental Disease, 198*(9), 691-694.

Propper, R. E., & Brunyé, T. T. (2013). Lateralized difference in tympanic membrane temperature: emotion and hemispheric activity. *Frontiers in Psychology, 4*.

Propper, R. E., Januszewski, A., Christman, S. D., & Brunyé, T. T. (2011). Increased anger is associated with increased hemispheric asymmetry: Support for anger-tympanic membrane relationships. *Journal of Nervous and Mental Disease*, 199(9), 716-20. doi:10.1097/NMD.0b013e318229d95a

Schmidt, L. A., & Trainor, L. J. (2001). Frontal brain electrical activity (EEG) distinguishes valence and intensity of musical emotions. Cognition & Emotion, 15(4), 487-500.

Sloboda, J. A. (1991). Music structure and emotional response: Some empirical findings. *Psychology of Music, 19*(2), 110-120.

Tatum, W. O. (2008). *Handbook of EEG Interpretation*. Demos Medical Publishing.

Tomarken, A. J., Davidson, R. J., Wheeler, R. E., & Doss, R. C. (1992). Individual differences in anterior brain asymmetry and fundamental dimensions of emotion. *Journal of Personality and Social Psychology*, 62(4), 676-687. doi:10.1037/0022-3514.62.4.676

Zimny, G. H., & Weidenfeller, E. W. (1963). Effects of music upon GSR and heart-rate. *The American Journal of Psychology, 76*(2), 311-314.

Zinner, L. R., Brodish, A. B., Devine, P. G., & Harmon-Jones, E. (2008). Anger and asymmetrical frontal cortical activity: Evidence for an anger–withdrawal relationship. *Cognition and Emotion, 22*(6), 1081-1093.

Navigating Embouchure Focal Dystonia: An Exploratory Case Study

Joseph Carucci

Abstract

Focal dystonia is a neurological disorder characterized by involuntary muscle contractions. In musicians, it is most often located in the arms, hands, face or lips (embouchure). Involuntary tremors occur when the musician is engaged in a musical task, resulting in challenges when performing music. Because there is no known cure for dystonia, musicians who have been diagnosed must search for ways to counterbalance the disorder.

The purpose of this study was to investigate whether an individual with severe embouchure focal dystonia could navigate the adjustment from performing on a brass to a woodwind instrument. The participant was a skilled trombonist who had battled for the latter part of his career with focal dystonia. After fighting with this condition, and ultimately surrendering his ability to perform on the trombone, he made a decision to embark on learning the saxophone. This exploratory case study documents a year of private saxophone lessons and the switch from one performance medium to another. Data was collected through observation, field notes, practice journals, and communication artifacts between the researcher and participant, and was coded for emerging themes.

Following a year of study, results indicate that the participant was able to redefine himself as a performing musician. His dystonia did not reappear with the use of a saxophone embouchure, and his identity as a saxophonist flourished. The results of this study offer musicians diagnosed with focal dystonia another option for maintaining a musical voice.

Keywords: focal dystonia, medical problems in musicians, embouchure disorder

Introduction

Focal dystonia is a neurological disorder characterized by involuntary muscle contractions. In musicians, it is most often located in the arms, hands, face or lips (embouchure). Involuntary tremors associated with dystonia occur when the musician is engaged in a musical task, and often make it impossible to continue creating music. Currently there is no known cure for dystonia, and treatments for symptoms are considered palliative (Lie-Nemeth, 2006).

Dystonia Medical Background

While the cause of dystonia is unknown, it is believed that it is a result of an abnormality or damage to the basal ganglia (Ward, 2006). This area of the brain is responsible for aspects of motor control and skill learning, and disorders in this region are characterized as hypokinetic (lack of movement) or hyperkinetic (excess of movement). Dystonia is also believed to affect the brain's neurotransmitters and sensory motor cortex, which impede the ability of different systems to communicate with each other. Damage to the basal ganglia and the sensory motor cortex will affect the way the brain processes information and generates a response, which results in an uncontrollable tremor (Hirata, Schulz, Altenmüller, Elbert, & Pantev, 2004; Ward, 2006). Many types of treatments are available for dystonia including Botulinum injections, medications, deep brain stimulus, surgery, and various other therapies (Lie-Nemeth, 2006; National Institute of Neurological Disorders & Stroke, 2015). Because little is known about the specific causes of dystonia, treatments for the disorder remain palliative rather than curative (Ward, 2006).

Dystonia in Musicians

Focal dystonia is not uncommon in musicians, with symptoms consisting of isolated involuntary tremors that only occur when the musician is engaged in a musical task. Seventy-five percent of those musicians diagnosed are male and develop dystonia around forty years of age (Lie-Nemeth, 2006). Of those diagnosed, pianists, guitarists, and brass players are the most common musicians affected (ibid). Symptoms in the hand include cramping, stiffness, and fatigue, while embouchure symptoms include embouchure tremor, involuntary lip movements, and jaw closure. Symptoms for musicians rarely involve pain. A brass player with embouchure focal dystonia may start with the symptoms in one octave, but over time the problems may become significantly worse (ibid).

Because musician's dystonia is task-specific and isolated to areas of the body that are used for music making, the early symptoms are often perceived as an issue with technique or insufficient preparation. Many musicians will intensify their practice sessions to compensate for the dystonic symptoms, which only exacerbates the situation.

Although no immediate cause for focal dystonia in musicians has been verified, tension and overuse may be a reason for developing the disorder. Compensating for a limiting physical attribute, such as smaller hands, may also be an element in acquiring the disorder. It is possible that psychological factors, such as stress, may also contribute. There tends to be a higher prevalence of anxiety and perfectionism in musicians with focal dystonia (Jabusch, Muller, & Altenmüller, 2004; Lie-Nemeth, 2006).

Treatments for Musicians

Because there is no cure for dystonia, there is no definitive treatment available for musicians. Most treatments aim to assist the musician in relearning the musical tasks without triggering the dystonia. Musicians suffering with dystonia have experimented with altering fingerings, postures, breathing techniques, and mental approaches, and had varying degrees of success. Medications, such as botulinum neurotoxin injections, have also been used as a treatment. These have also had varying success rates and are mostly limited to those suffering from symptoms in their hand or arm (National Institute of Neurological Disorders & Stroke, 2015). Embouchure focal dystonia is much harder to treat (Jabusch & Altenmüller, 2006). Bromocriptine has been experimented with as a possible medication, but it is usually recommended that musicians try to modify their mouthpiece and retrain their embouchure (Lie-Nemeth, 2006). It is suggested that focal dystonia patients avoid dramatic increases of practice time, speed, or intensity, and place a priority on reducing stress, tension, and anxiety. Using appropriate warm-ups and cool-downs, taking breaks, and maintaining good posture and body conditioning are also important in managing the disorder (ibid).

Unfortunately, even when action is taken to remedy the effects of focal dystonia, there are individuals who are unable to fully resume their pre-dystonia performance abilities (Guptill, 2011). This can result in a loss of musical identity, depression or anxiety for the musician. One possibility for individuals suffering from focal dystonia is to explore the option of pursuing a new performance medium. To date, there are no studies that have documented the process of learning a new instrument for musicians with focal dystonia. The purpose of the current case study was to explore this process with a professional musician that has suffered debilitating embouchure focal dystonia.

The Development of Richard's Dystonia – A Vignette

Focal dystonia first appeared in "Richard's" lips at the age of forty-three. Richard was already established as a professional trombonist and served as a trombone professor at a major university in the United States. The tremor in his lips appeared suddenly while teaching a private lesson and was concentrated in the middle register of the trombone.

The dystonia was devastating to Richard, as his profession was to perform and teach music. He sought treatment and advice from professional musicians, former teachers, and colleagues throughout the United States including formal treatment from a specialist in retraining musicians diagnosed with the focal task-specific dystonia.

During his treatment, Richard focused on performance techniques that would retrain his brain to minimize the dystonic effects. Ultimately, he changed his approach to using 'effortless' air and focusing on the concept of tone as opposed to the physical technique of creating a tone. Improvements were apparent when trying to overcome the dystonia, but the tremor never completely subsided. He continued to perform on a limited basis and teach for approximately fifteen more years, until retiring. Although the dystonia negatively affected his ability to perform, he believed it strengthened his teaching approach.

Upon retirement, Richard gave up performing on the trombone entirely. After a long performance hiatus and without the expectations of performing at a professional level, he attempted to play the trombone again. However, the dystonia was still apparent. At this time, Richard made the decision to give up the instrument, sold most of his trombone equipment, and decided to embark on the process of learning a new instrument – the tenor saxophone. Richard was drawn to this instrument because of the similarities in range and timbre to his previous instrument, the tenor trombone. He felt the tenor saxophone would offer him a similar musical voice to what he had experienced in his four decades of trombone performance.

The Research Study

In order to begin a new instrument, Richard knew he would need to begin taking private lessons on the tenor saxophone. He was referred to the researcher, who is both a saxophonist and a music educator that is familiar with teaching adults. Both agreed it would be worth documenting the process of navigating the challenges of focal dystonia while learning a new instrument. Appropriate IRB approval was attained, and the research study commenced.

A case study methodology was employed, which allowed the researcher to work closely over time with the participant and to collect a variety of data within a specific context (Creswell, 2007; Zainal, 2007). The study was descriptive in its explanation of the real-life context in which it occurred, and was bound by time and place, as well as a final goal (Baxter & Jack, 2008; Creswell, 2007) of a public performance.

The participant's progress and experiences learning the tenor saxophone were documented over the course of thirteen months. During that time, the subject took sixteen private individual saxophone lessons (each of which were two to three hours long) with the researcher. Data was collected through observations, field notes, practice journals, musical recordings, and communication artifacts between the researcher and participant.

In order to track the musical progress during the time between lessons,

Richard kept a practice journal. The journal contained information and observations that the participant had regarding his experience and was shared with the researcher. Questions regarding saxophone techniques, equipment, and practicing were also noted. Over two hundred emails were exchanged with the researcher during the data collection period, further documenting Richard's progress, questions, and interests. All data was eventually coded for emerging themes, and informant feedback ensured the accuracy of the information being collected.

Themes of Richard's Saxophone Journey

Richard had already established a strong musical base over the course of his forty-year career as a professional trombonist. This allowed him to implement many advanced musical elements much quicker than the average beginning saxophonist. This can be observed in several stages of his progress over the first year of lessons.

Vibrato

The participant began to experiment with adding vibrato to his lyrical playing in the fourth month of studying the saxophone. Not only is this much earlier than a typical saxophonist would begin vibrato, but he also had a concept established on how he would like it to sound based on his previous musical experience. Although he spent a short time during one lesson being sure that his technique for creating vibrato was correct (using jaw vibrato), this element of his saxophone playing was implemented almost entirely by his motivation to create a more lyrical phrase.

Articulation

Articulating and the effects of accents, staccato, legato, etc., on a woodwind are very different from those on a brass instrument. Adjusting to articulating on a reed was a challenge for Richard. In both lesson observations and throughout his practice journal, there are notes regarding the experimentation of various syllables to create musical effects. During the first four months he often referred back to breath attacks, which he used regularly on trombone.

Although the subject struggled with articulation throughout almost the entire first year of studying the saxophone, his experimentation and persistence eventually led him to be adaptive in various styles of articulation. At one point, he indicated, "I'm working hard on articulation as well to make it more natural; I'm trying to do this by ear (concept) rather than thinking about HOW to do it. I'm making up some exercises for this." During the year he spent considerable lesson time transcribing articulations from the demonstration and play-a-long CD that accompanies the book *15 Easy Jazz, Blues & Funk Etudes* (Mintzer, 2000), which eventually led to much more versatility in his articulation.

Facility

Building smooth and fast facility on the saxophone was one of the biggest challenges. The use of dozens of fingerings to create pitches as opposed to seven slide positions is a big adjustment. This is coupled with the necessity to read music in treble clef when the subject has spent most of his career performing music in bass clef posed a challenge. By concentrating on articulation, and saxophone facility, the focus was diverted away from the focal dystonia and onto personal improvement.

Identity

As the year progressed, Richard identified himself more as a saxophonist than as a former trombonist. He expressed his personal excitement in being able to play music again:

> I truly believe that each of us has a purpose in life and something that is really designed to give us joy, and for me, that is expressing myself through a musical instrument that has meaning to me. There is a part of [a person's] soul which can never be expressed in words but only through music. I have missed that for over 2 years, and now I have a chance to get that back again. Each day I am filled with incredible hope and joy to do that once more! I absolutely LOVE practicing. In fact, I have a hard time putting the horn down. Sunday's I usually take off because I think it keeps me fresher and more into the instrument (sometimes I play anyway

because I just really want to), but today I was just dying to play my sax again.

In addition, he made a monetary commitment by investing in saxophone books, recordings, and equipment. He invested in a new mouthpiece that would help him create a tone closer to the concept that he was striving for, and eventually began acquiring new books because they were intriguing to him, and not just because of his teacher's recommendations.

Practice Habits

The participant began his saxophone experiences with strong practice habits and motivation. Even in the early stages, Richard was consistently practicing several hours a day, noting "I have to force myself to quit." During the first few months he would ask the researcher to help him create a practice routine. The longer he took lessons, he would simply outline his practice routine himself for my reference. After several months of lessons, he began to record various solos and duets during our lesson time to utilize when he was practicing at home. He continually logged his practice routines and shared these with the researcher for comments.

Musical Goals

Throughout the year, Richard's musical goals became clearer. After four months of study, he outlined his long-term and short-term goals, with performances being a long-term goal, and articulation and effortless high notes a focus for the short-term. Early on he noted, "I keep having to take the long view on this learning process...I am determined as long as the focal dystonia doesn't get in the way."

Richard's organized and focused practice method helped him reach his short and long-term goals. His high notes improved enough that he added the altissimo register to his saxophone range, and only thirteen months after starting saxophone he began performing in church and at a nursing home. He has continued to prepare music for performances and plans to continue studying the saxophone and increasing his performance level.

Tone Quality

By following Richard's progress over the course of the year, several important observations regarding his embouchure focal dystonia can be made. First, the focal dystonia was apparent in the early stages of the subject's saxophone study. Since both he and the researcher had the background to understand that the brain needed to create new neurological pathways to overcome the tremor in his lips, the focus remained predominantly on tone production and musicality. Tone quality was a main focus for Richard during his study:

> I think it's so important to have a great sound model in my head from the beginning or I'm going to be using my own concept, which might not be that great or refined, and could thus negatively affect my production... I'm just trying to start this process with the right sound in my head; and if I don't do that, I think I'm putting the cart before the horse.

When trying to overcome focal dystonia earlier in his career, he had developed a practice of focusing on sound and musical concept to avoid any unnecessary tension in his aperture. Even though Richard had switched to an embouchure in learning to play the saxophone, that same approach was still necessary with the tenor saxophone. He experimented with various reeds and mouthpieces, and a lighter reed created the least resistance leading to a more relaxed approach to tone production. He also chose a mouthpiece based on easier tone production, and later paired a new mouthpiece with a lighter reed to relieve tension.

Richard's tone would still tremor from the dystonia for the first five months of study, but that was never a focus of conversations in his lessons. He experimented with embouchures until he found one that created the least amount of tension. Gradually the dystonia subsided and halfway through the year, there was no indication of the tremor in his playing. Shortly thereafter, Richard stated, "I don't think focal dystonia will be an issue at all with this instrument."

After just over one year of learning the saxophone, the participant has been able to perform music again without the effects of the dystonia. He has also begun to redefine himself as a musician, transcribing solos and choosing literature to prepare for future performances. At thirteen months of study, he was able to perform for a public audience, which was a milestone in his learning process. The results of this study hope to offer musicians diagnosed with focal dystonia another option to maintaining a musical voice.

Conclusion

Learning a new instrument has been a positive experience for the participant in this study. He has been able to apply the strong musical skills that he acquired on the trombone to his saxophone performance. In the course of approximately one year, he has been able to perform music again without the devastating effects of the embouchure focal dystonia.

Since the dystonic symptoms were apparent when he began to study the saxophone, some adaptations were necessary to avoid the disorder or to keep it from progressing. Most importantly, a conscious effort was made to make tone production on the saxophone as tension-free as possible. These included using a lighter-strength reed (2.5) and a mouthpiece with a smaller tip opening (Selmer C*). It was also important to use an embouchure with the least amount of tension as possible. This meant adopting an embouchure with the bottom lip folded out, which creates equal pressure around the mouthpiece. Articulation and facility on the saxophone were the two biggest technical struggles for the participant. Focusing on these two aspects of saxophone playing served as a distraction from the participant's struggle with focal dystonia.

While by nature case studies are not generalizable, Richard's experience in finding a musical voice on another instrument offers promise for other musicians suffering from embouchure focal dystonia. Further research is certainly warranted to help provide additional information on this unique change of performance mediums. A longitudinal follow up on this participant is recommended to determine if the dystonic symptoms continue to be a non-factor in his performing on the saxophone. In addition, it would be worth noting if any other setbacks may arise with increased experience on the instrument. For musicians who struggle with focal dystonia, the goal of shifting the treatment of dystonia from palliative to curative is a worthy cause.

References

Baxter, P., Jack, S. (2008). Qualitative case study methodology: Study design and implementation for novice researchers. *The Qualitative Report, 13*(4), 544-559.

Creswell, J.W. (2007). *Qualitative Inquiry & Research Design: Choosing among five approaches.* Thousand Oaks, CA: Sage Publications, Inc.

Guptill, C.A. (2011). The lived experience of professional musicians with playing-related injuries: A phenomenological inquiry. *Medical Problems of Performing Artists, 26*(2): 84-95.

Hirata,Y., Schulz, M., Altenmüller, E., Elbert, T., & Pantev, C. (2004). Sensory mapping of lip representation in brass musicians with embouchure dystonia. *Neuroreport, 15*(5*),* 815–8.

Jabusch, H. C., & Altenmüller, E. (2006). Focal dystonia in musicians: From phenomenology to therapy. *Advances in Cognitive Psychology, 2*(2-3), 207-220.

Jabusch, H.C., Muller, S.V., Altenmuller, E. (2004). Anxiety in musicians with focal dystonia and those with chronic pain. *Movement Disorders, 19*(10), 1169-75.

Lie-Nemeth, T. J. (2006). Focal dystonia in musicians. *Physical Medicine and Rehabilitation Clinics of North America, 17*(4), 781-787.

Mintzer, R. (2000). *Fifteen Easy Jazz, Blues and Funk Etudes*. Emeryville, CA: Alfred Music.

National Institute of Neurological Disorders and Stroke (2015, July 6). Dystonia fact sheet. Retrieved from http://www.ninds.nih.gov/disorders/dystonias/detail_dystonias.htm

Ward, J. (2006). *The Student's Guide to Cognitive Neuroscience*. New York, NY: Psychology Press.

Zainal, Z. (2007). Case study as a research method. *Jurnal Kemansiaan, 9,* 1-6.

A Correlational Study of Performance Anxiety of Non-Musician Performing Musical Tasks

Derrick Alan Crow, Ph.D.

Abstract

Performance anxiety is a common problem among musicians, can cause severe and debilitating effects, but is also present in non-musicians. The purpose of the experiment was to test the effect of performance stressors on the anxiety levels of non-musicians. All participants were from a Midwestern university, and had received a basic level of sight-reading, singing, and keyboard skills. The participants were asked to complete four musical tasks. After the completion of those musical tasks, the participants were asked to complete a survey, which explored perceived stress and perceived performance success. All performances were recorded and evaluated by two judges. All performances were recorded and evaluated by two judges. The judges looked at performance quality and observable performance anxiety markers.

The initial hypothesis was that the presence of an audience would correlate with the experimental group reporting a higher level of perceived anxiety. This assumption was not supported by the difference between experimental and control samples.

Introduction

Performance anxiety is a common problem among musicians. It can cause severe and debilitating effects but is also present in non-musicians. The purpose of the experiment was to test the effect of performance stressors on the anxiety levels of non-musicians, and to closely examine the connection between the presence of an audience and perceived anxiety.

Across the United States, students are asked to sing and play instruments in front of their peers and teachers. Many of them have severe performance anxiety that is caused by the task we are asking them to do – or is it the fact that they feel we stand in judgment?

The purpose of this study was to examine the audience effect on perceived anxiety. The researcher used mixed methodology of self-reporting, behavioral observations, and adjudicated performances. The data will be evaluated for relationships between audience and perceived anxiety, perceived success and perceived anxiety, and perceived success and audience effect. The researcher examined how each task affected the participant, and how an audience augmented or negated this effect. The achievement of each participant was measured by two judges, and a behaviorist observed each performance for physical markers for performance anxiety. These measures provided a rich depth of data for analysis.

If the presence of an audience is found to have an effect on perceived performance anxiety, such a finding might offer insight into dealing with performance anxiety in students and young performers. Often, the lack of performance success can be linked with an increased level of performance anxiety, and through better understanding, the instructor can develop behavior modification programs to aid all students perform with less anxiety. Ostwald, Avery, and Deschamps (1991) suggest that prolonged stress can cause long-lasting medical problems, which is why understanding the effect of an audience on performance anxiety is important.

Literature Review

Anxiety has been defined as either state or trait. State anxiety is caused by situational circumstances that increase your anxiety, while trait anxiety is an internal level of anxiety. Kemp (1996) offered the idea that musicians were predisposed to trait anxiety, and suggested that the more personal responsibility a musician takes for their performances, the more likely the participant is to have higher levels of anxiety. Kokotsaki and Davidson (2003) state that even though performers are more likely to have higher trait anxiety, they are still affected by state anxiety. In this study we will examine the effects of state anxiety by replicating the study by Leblanc, Jin, Obert, and Siivola (1997).

In the Leblanc, Jin, Obert, and Siivola (1997) study, the researchers sought to examine the effect of an audience on high school band members. They found that the older and more trained the participant, the greater the performance anxiety. Their study also suggested that the size and membership of the audience directly correlates with the perceived anxiety of the participants.

Wan and Huon (2005) believe anxiety comes from internal ego issues, and that the state can be influenced by rewards, punishments, and the presence of adjudicators. Comparing these two studies suggests that the effect of external stressors on our perceived anxiety is related to the subject's expectations. Kemp (1996) believes the musicians committed to music have difficulty separating personal identity from their skill as musicians, which shows how personal expectations can increase state anxiety.

While musicians with higher levels of formal training are subject to more self-imposed stressors, Kokotsaki and Davidson (2003) believe that if the pattern becomes commonplace, the performer can become desensitized to the stressor. While this belief seems to run counter to the previous ideas, it in fact presents the variable of familiarity with task. In the simplest terms, the more you do an action, the more you become conditioned to it.

Methodology

The participants (*n*=22, 6 males and 16 females) ranged in age from 18-22, were recently enrolled in a music education course for elementary education majors, and had received basic training in sight-reading, singing, and keyboard skills. The subjects were volunteers from the undergraduate elementary education program at a Midwestern university, and had limited exposure to musical performance. The population was composed of elementary education, early childhood, and special education majors. The sex of the participants was predominantly female, which is representative of these majors. The participants were divided into control (*n*=11) and treatment (*n*=11) groups. The control group was asked to perform all tasks in front of the researcher, and the subjects assigned to the treatment group were asked to perform all tasks in front of both the researcher and an audience of three observers. All subjects' performances were recorded for later performance and behavioral analysis.

Treatment Conditions

The audience was composed of three graduate students pursuing terminal degrees, who had served as associate instructors, and who had taught music for 8 years. The audience members were dressed professionally and conducted themselves as attentive listeners. They were seated less than 8 feet from subjects during their performances with only the piano (placed at a 60-degree angle) separating the performer from the audience.

Audience

The audience for treatment group was made up of graduate students (*N*=3) with no direct connection to the subjects. These audience members did not need experience with objective analysis of young performers, but were asked to appear to be making notes on each performance.

Judges

Two judges reviewed the recordings. The judges varied in age and experience. The first judge had eight years of public school teaching and two years of college teaching experience, and had taught and evaluated performance skills of a similar population as the participants. The second judge had four years of public school teaching and six years of college teaching experience. Both judges share similar experiences, and both had evaluated students of comparable skills to the sample. The judges used a criterion-based rubric to evaluate the performance. The rubric addressed pitch, rhythmic accuracy, ease of performance, and tone quality.

Procedure

The full sample (*n*=22) will complete the same four musical tasks. The participants were given two minutes to prepare a Do Re Mi composition, but were not required to use the full time. The same composition was used for each performance task. First, they performed composition of the keyboard and sang: second, the participants performed on the keyboard without singing; third, the participants sang the composition without a keyboard; and fourth, they sang with a recording. After completing each performance, the subjects filled out a survey addressing perceived anxiety and performance success.

Results

The results for subject responses were run for the full sample and for both control and experimental groups. Descriptive data were established for all subject groups and as a complete set. The survey data were then examined for correlational relations, and then compared within and between groups.

The descriptive data for all questions and all groups were: the mean for the survey questions for the full sample (*n*=22) were between 2.56—4.53 for all items, with a range of 4—6, and a standard deviation between 2.76—4.71. The control group (*n=11*) had means between 2.73—4.71, with a range between 3—6, and standard deviations between .76—2.22. The Experimental group (*n=11*) had means between 2.5—4.5, with a range between 4—6, and standard deviations between 1.51—2.27. After establishing the descriptive data, the researcher examined the correlation between the survey questions.

The survey questions were examined for internal reliability. It was found that the six questions could be collapsed into two categories. The first category can be referred to as perceived performance success, consists of Questions 1, 2, 4, and 5, and was found to have a .96 Cronbach's Alpha Coefficient score, which suggests a strong internal relationship between these questions. The second category will be referred to as performance anxiety and consisted of Questions 3 and 6. These questions were found to have a .93 Cronbach's Alpha Coefficient score. The questions were collapsed into 2 categories and were found to have a highly significant correlation (p> .01) between categories. The composite measures were then compared across each of the four tasks to examine any correlation between tasks. (See Table 1)

No significant correlation was found between groups, but a larger standard deviation was observed in the experimental group. There was also a slight difference in the cumulative perceived stress between groups and showed a slightly higher mean reported by the experimental group.

The judges then evaluated the recordings. The judges examined the performances, and looked at rhythm accuracy, pitch accuracy, intonation, and performance quality. The judges' scores demonstrated an internal reliability of .901 or higher Cronbach's Alpha Coefficient scores for each task, and a Cronbach's Alpha Coefficient score .890 across all tasks. Judge 1 reported a mean of 26.69 with a standard deviation of 1.30, and Judge 2 reported a mean of 24.50 with a standard deviation of 4.26. Accepting this reliability, the judges' evaluations were compared

Table 1. A Pearson Product Moment Correlation of Perceived Stress and Perceived Performance Success Variables

	T1 Per	T2 Per	T3 Per	T4 Per	T1 Stress	T2 Stress	T3 Stress	T4 Stress
T1 Per		.813**	.602*	.718**	.622**	.670**	.525*	.718**
T2 Per			.497	.657**	.657**	.891**	.555*	.761**
T3 Per				.851**	.312	.284	.683**	.638*
T4 Per					.337	.454	.508	.506*
T1 Stress						.763**	.744**	.636*
T2 Stress							.562*	.739**
T3 Stress								.741**
T4 Stress								

against performance success and perceived anxiety. No significant correlation was detected.

Discussion of Findings

The researcher's hypothesis that the presence of an audience would correlate with the experimental group reporting a higher level of perceived stress was not supported by the sample. The difference between means was non-significant, but the standard of deviation in each task was larger in the experimental group. There could be several explanations for these results. First, the population of the study, unlike trained musicians, may be able to separate themselves from the experience. This detachment would allow the participants to perform the tasks without fear of failure. The second possible reason for non-significant results could be the testing area. The piano in the room was an upright and allowed the subject to obstruct their view of the audience. This could also account for the larger variance in the standard deviation; some subjects may have had more success obscuring the audience. The final possibility is that the performance tasks did not induce anxiety.

The correlation between the judges and the subjects' self-reports showed a non-significant correlation. This presents the question: do untrained performers view musical performance differently than the judges? When the performance success scores were compared, the participants' responses showed a lower mean than those of the judges. This could imply that even though the subjects reported relatively low feelings of anxiety, their low perception of performance success may have been a symptom of unrealized anxiety. This hypothesis is supported by the significant relationship between participant-reported perceived anxiety and performance success, which suggests these two variables are closely related in the minds of the participants.

This hypothesis is strengthened by the behavioral cues observed by the researcher. All but one participant in the control group used the whole time allotted to practice the composition, but eight subjects in the experimental group did not. Only three used the full two minutes, and the remaining eight participants asked if they could begin the tasks less than 40 seconds into the practice time. Of the three participants who did use the full time, two practiced, but the third sat at the piano and stared at the music. The participant did not move or make any clear signal that they were working on the piece. This subject during performance did demonstrate other signs of performance anxiety. Her voice was weak and wavering, and demonstrated muscle tension, which as the task progressed could be seen to increase. The tension started in the subject's hand and extends to the neck. The one subject who did not exhibit visible stressors demonstrated a higher level of musical skills.

The performance totals and stress totals correlate across tasks and suggest that if the subject were successful on one task, they would have success in the other tasks. Only when asked to sing without aid of pitch reference from keyboard or recording did success in other task not predict success. Through musical training many performers develop relative pitch, which allows them to maintain a sense of tonal center, and the freedom to sing with continual pitch reference.

The implied trends present an interesting view at non-musicians' performing performance tasks. The data, while interesting and thought provoking, does not offer a concise picture of the influences on performance anxiety of non-musicians, but it does offer some possibilities. The item that I feel deserves attention is the ability to disconnect from musical performance. Musicians define their identity by their prowess in performance, and as Kemp (1996) suggests, musicians have a high level of anxiety due to fear of failure. The ability to have a separation between performance and self is a skill that could assist musicians, and possibly prevent people from leaving the profession due to anxiety issues. This suggests that performing is not innately stressful, but it is our letting our profession become our identity, and our desire to compete, that it becomes a source of anxiety. If this proves to be true and is found to be generalizable, then we should reevaluate our approach to teaching music, from the kindergarten to the graduate level.

The second trend, which became clear, was perception – the perception of participants and of the judges. The subjects' responses show a lower mean than the judges' evaluation of performance. The limited experience could have contributed to their critical evaluation of their performance, or the correlation between performance success and perceived anxiety could suggest that their feeling of poor performance could be a result of anxiety. Between these two trends, I am left asking: are non-performer more detached and experience less anxiety, or do their feelings of performance success represent a feeling of performance anxiety?

The study, while presenting more data on an important topic, needs refinement. The first steps are to remove confounding variables of composition difficulty, having an unobstructed view of the audience, having a larger audience, and having a live pianist to add an additional task - to examine the question more clearly. Increasing the difficulty of the composition should raise the overall anxiety level. Increased difficulty of performance could also be accomplished by eliminating the preparation time, but with this population, true sight-reading is not possible, and could become a confounding variable in and of itself. The increased difficulty of the composition could make it accessible by the population, but would require a higher level of skills, and might amplify any audience effect on perceived anxiety in the treatment group. Changing the piano to a Clavinova or a grand piano (with the lid shut) would create an unobstructed view of the audience members. If the size of the audience was increased, the effect of the audience members could be amplified.

The research did not find a significant correlation between any of the original questions. The significant relationship found suggests that the population does relate the feeling of anxiety with task success, and with this information, future replications could give a clearer understanding of performance anxiety in non-performers.

References

LeBlanc, A., Jin, Y., Obert, M., Siivola, C. (1997) Effect of Audience on Music Performance Anxiety. *Journal of Research in Music Education.* 45.3.480-496

Kemp, A. (1996) *The Musical Temperament: Psychology and Personality of Musicians*. New York, New York. Oxford University Press

Kokotsaki, D., Davidson, J. (2003) Investigating Musical Performance Anxiety among Music College Singing Students: A Quantitative Analysis. *Music Education Research*. 5.1.45-59

Ostwald, P., Avery, M., Deschamps, L. (1991) Musculoskeletal Problems in Instrumental Musicians. *Performing Arts Medicine.* New York, New York. New York Press

Wan, C., Huon, F. (2005) Performance Degradation Under Pressure in Music: An Examination of Attentional Processes. *Psychology of Music*. 33.2.155-172

Wilson, G. (1997) Performing Anxiety in: D.J. Hardgraves & A. C. North (Eds). *The Social Psychology of Music.* New York, New York. Oxford Press 229-243

A New Holistic Approach for Healing with Music: Love Based, Positive Lyrics with Improvised Music

Carla Rose Kelly & Julia MacKinley

Abstract

This research outlines two main foundational components of a new holistic approach to using improvised music as a tool for healing anxiety, depression and pain. The holistic method includes mind, body, and spiritual considerations. The first foundation is based on the spiritual teachings that all healing comes from love. Therefore, the practitioner uses a mindful and meditative approach by ensuring every breath is from a place of love, and the mind is 100% focused for the duration of the music healing session. Some of the latest research on mindfulness combined with neuroscience will be presented: "Where attention goes, neuro-filing flows, and neuro-connection grows."

The mind-body connection will be discussed through modern research in multi-disciplinary fields, bridging neuroscience, psychology, neurotheology, music, and medicine. This introduces the second foundational aspect of this new methodology in the exclusive use of positive words for lyrics. Modern research in neuroscience and psychology show how words can change our brain and influence our behavior and emotions. One important brain area involved in this process is the amygdala - the area involved in processing of emotional stimuli, memorizing these reactions and activating responses such as anger. Deficits in amygdala function have been associated with depression, anxiety and post-traumatic stress disorder among many others. This important brain area is affected when we are exposed to negative words. The researchers will also present findings from their qualitative and quantitative data in regard to 120 participants exposed to positive affirmations for a set of three one-hour music sessions with significant results. This experiential workshop will offer participants an opportunity to experience as well as participate in the mindful focus of a love-based breath as well as using positive affirmations with music. Discussion will follow sharing feedback on participant experiences with words, music and emotions as well as any awareness of affects in the mind, body and spirit.

Keywords: research on holistic healing with music, research on positive lyrics, research on love-based music, mind-body-soul research for healing with music, research on improvised music with positive lyrics, research on spirituality and music, research on music for healing. Mind-body-spirit research for music, Sacred Scale Music research

Introduction

Music has been used as a form of medicine since the beginning of time, but the field of holistic healing with music has not previously been studied scientifically. Holistic healing approaches consider the three aspects of being human: the mind, body and soul. There is substantial research on the effects of music on the mind and body with little integration of the spiritual aspect as this is very complicated to measure scientifically. The goal of the current paper is to integrate the known research on the mind-body connection with music to the long known, but understudied, spiritual connection. In order to begin formulating a new holistic approach connecting the spiritual to the known mind-body research, spiritual teachings will be used that are accepted around the world. Despite the differences in world religions, they all agree on some basic principles. Let's consider these two spiritual beliefs:

1) Words are our God/creative power
2) All healing comes from love

Spiritual Foundation #1: Words are our God/creative power

This spiritual foundation can be found across various religions and philosophical teachings. From the Gospel of St. John in the Christian Bible, it reads: *"In the beginning was the WORD, and the WORD was God, and with the WORD was God, and the WORD is God."*

In the Hindu Vedas, it is stated: "In the beginning was Brahman with whom was the Word. And the Word is Brahman." In the Mayan text, *Popol Vuh*, the first humans were given life solely through the power of the word. The ancient Egyptians believed that the God Thoth created the world by his voice alone. The Hopi Indians tell the story of Spider Woman, who sang the Song of Creation over the inanimate forms on earth, bringing them all to life. As can be seen through these examples, very similar beliefs of words being our God or creative power are held across religions and cultures.

If our *words,* which begin with our thoughts, are our "God/Creative" power over others and ourselves, then it is essential that we understand how words affect the mind and body. This will be presented later in this paper. Let us continue with other spiritual ideals to understand the new holistic approach to using music for healing.

In considering our spirituality, we could view ourselves as a soul having a physical experience. The body could be equated to the "temple" that the soul lives in, and the mind the "master control" for all its operations. If we take into consideration the spiritual teachings that we are all made "in the likeness of God", then is the soul within us a "God in the making"? If we also take into consideration that the WORD is our God or creative power, is it through the use of our WORDS that we create "good" or "evil" in ourselves and others? Is the purpose of our physical existence to find our potential, and eventually reach our fully actualized selves or God selves? Does the "temple" (body) or "master control" (mind) for our soul break down when we are off the path to this actualization? Are our thoughts and words responsible then for our own well-being? What impact do our thoughts and words have on others' well-being? Research presented in this paper will explore these issues.

Spiritual Foundation #2: All healing comes from love

If we accept that all healing comes from love, then it is essential that this be the starting place for healing with music. The practitioner of this new holistic approach to healing with music will center himself or herself in a place of love and conduct the methodology from this place with 100% mindfulness. The method and reasons for this will become obvious with the research detailed in this report.

If we were to accept the above statements about our spirituality as the truth for our physical existence, then we could all be working towards the pureness and unconditional qualities of love that we imagine God symbolizes. If we need to heal all things within ourselves to reach our fully actualized self, then the key would be to start from a place of love, grow that love within ourselves, until we are eventually the pure essence of love that a God would symbolize. During this process of "actualization" or "working towards our God selves", all thinking, words, and actions would need to be from a place of pure love. If the "God essence" (soul) within us is "creating" from a place of love, then the "temple" (body) and "master control" (mind) should be integrated and healthy. Let's look at the thoughts and research from some of the greatest minds bridging their findings together to support this new holistic approach to healing with music.

Philosophical Background

Greek mythology ascribed to music a divine origin and named its inventors, "Gods". They understood that music had magic powers. It could heal sickness, purify the body and mind, and work miracles in the realm of nature. The Greeks achieved insights and formulated principles of the understanding of music, many of which have not been superseded to this day. Their understanding of the nature of music, its place in the cosmos, its effects and its ethically correct uses in human society are profound. Pythagoras, Plato, and Aristotle were some of these Greek individuals, who understood the power in music. They taught the world to understand that the whole spiritual and physical universe was the understanding of numbers. So the system of music, sounds, and rhythms, being ordered by numbers, were conceived as exemplifying the harmony of the cosmos and corresponding to it.

Pythagoras, in his "Doctrine of Ethos," stated

> *"music is a microcosm, a system of sounds and rhythms ruled by the same mathematical laws that operate in the whole of the visible and invisible creation. Music was not only a passive image of the orderly system of the universe, it was also a force that could affect the universe." (p. 7)*

The word *ethos* in this context means "moral character", and Greek ethos theory concerns the ways in which music can convey, foster, and even generate ethical states.

Plato, in his "Music of the Spheres," states:

> *"There is a power in music akin to the power of words for influencing human thought and action, and that therefore, an artist, whether in music or words, is under obligation to exercise this power with due respect for its effect on others". (p. 7)*

Aristotle, in his "Doctrine of Imitation," states:

"Music directly imitates (represents) the passions or states of the soul: gentleness, anger, courage, temperance and their opposites and other qualities. Hence, when one listens to music that imitates certain passions, he becomes imbued with the same passion and if over a long time, he habitually listens to the kind of music that rouses ignoble passions, his whole character will be shaped to an ignoble form." (p. 7)

It is important to note, that this philosophical background is also our earliest known record of music from the book The History of Western Music (Grout, 1980). The power of words and music was known in 500 BC by these philosophers to shape our personalities, values, morals, character and behavior. Music was also understood to be our connection to something bigger than our selves … the universe or God. Perhaps this is why music has always been an integral part of church and ceremonial services. Modern research is proving their teachings to be true.

Modern Research on the Power of Words

The following studies bridge multi-disciplinary fields into one framework to underlie the new holistic approach to healing with music. These fields include psychology, neurobiology, neurotheology and neuroscience.

To best understand why positive lyrics in music is so critical, it is important to understand the impact of negative lyrics on one's brain. Within the last 10 years there has been alarming research conducted on the use of violent words in music. The American Psychological Association released a report on the effects of violent song lyrics on aggressive thoughts and feelings (Anderson & Carnagey, 2003). This was a groundbreaking study due to the popular opinion that listening to violent music could be a cathartic experience. However, the results of the study showed just the opposite. Even violent songs that were performed in a humorous or satirical way lead to an increase in hostile thoughts and feelings. In one substudy, a series of violent and non-violent songs were sung by the same artist and measures of aggressive thoughts and feelings were administered afterwards. One of these measures involved classifying words that could have both aggressive and nonaggressive meanings (i.e. rock and stick). Results of the experiment showed that violent songs lead to more aggressive interpretations of ambiguous words. It was explained that this could lead to real-world implications if listening to violent music can "influence perceptions of ongoing social interactions, coloring them with an aggressive tint". By viewing the world through an aggressive lens, it can lead individuals to instigate a more aggressive response than if one had not been experiencing such an aggression-bias.

Dr. Fields, a neurobiologist, published a book titled *The New Brain* which claims that verbal abuse causes far more than just emotional harm. He cited changes in the corpus callosum in individuals exposed to verbal abuse. The corpus callosum is a bundle of nerve tissue containing over 200 million axons (nerve fibers that carry electrical impulses from neurons' cell bodies) that facilitates communication between the two sides of the brain. The corpus callosum is the largest collection of white matter within the brain, and it has a high myelin content. Myelin is a fatty, protective coating around nerves that facilitates quicker transmission of information. If this is damaged from continuous exposure to negative words, the integration of brain parts will be impaired.

Further research in the field of neurotheology presented in the book *Words Can Change Your Brain* (Newberg & Waldman, 2012) shows the links between spirituality and brain changes. Their research involved MRI scans while the subject was exposed to negative words. They found that "a single word has the power to influence the expression of genes that regulate physical and emotional stress". Just by showing participants the word "NO" for less than one second, they saw a sudden release of stress-producing hormones and neurotransmitters. These chemicals impede the normal functioning of the brain and can lead to impaired logic and reasoning abilities, impaired language processing and communication skills. This finding was exacerbated in highly anxious and depressed people. It was also found that the longer people ruminated on the negative words, the more damage they experienced in terms of ability to regulate memory, feelings and emotions

(Talarrovicova et al, 2007). Adding emotional intensity to the negative words created further problems. "If you vocalize your negativity, or even slightly frown when you say "no", more stress chemicals will be released, not only in your brain, but in the listener's brain as well. The listener will experience increased anxiety and irritability, thus undermining cooperation/teamwork and trust" (Tessitore et al., 2002).

Research on popular music has explored its effects on schoolwork, social interactions, mood and affect, and particularly behavior. The effect that popular music has on children's and adolescents' behavior and emotions is of paramount concern according to the American Pediatric Association who compiled 106 research papers regarding the same. Lyrics have become more explicit in their references to drugs, sex, and violence over the years, particularly in certain genres. A teenager's preference for certain types of music could be correlated or associated with certain behaviors. As with popular music, the perception and the effect of music-video messages are important, because research has reported that exposure to violence, sexual messages, sexual stereotypes, and use of substances of abuse in music videos might produce significant changes in behaviors and attitudes of young viewers. The imaging coupled with the messaging of the words watched repetitively is shaping the values, and morals, shaping the personality, affecting the behaviors of the viewers.

This modern-day research from multi-disciplinary fields conclusively supports the early philosophers' teachings of Plato, Aristotle and Pythagoras (500 BC) on how music and words affect our personality, behavior, morals and values. It is also interesting to note the damage to the brain with repeated exposure to negative words. If the brain, which is the master control for all of the body's operations, is not functioning at optimal level, how can it manage the body which is the temple in which the soul lives?

How does the brain process music and affect emotion?

Music affects the body and brain in many positive ways; however, until recently, the way in which music is processed by the brain has not been well understood. One hotly debated question in the field of music research has been whether the brain processes music as one entity, or whether the lyrics and music are processed separately. Interest into this topic began when it was discovered that individuals with aphasia, a disorder involving the inability to produce or comprehend language, could still hum a tune. This suggests that the words and the music are, at least in part, processed separately.

Further research was conducted to look at whether the music and the words are processed in the same area and it was hypothesized that there are two main systems involved in the processing of music in the brain. One system in the temporal lobe, an area responsible for memory, appears to be activated during familiar melodies or songs. The other system in the frontal lobe is activated when hearing new music and helps us follow the rules that are important to both music and language (such as syntax or rules of harmony). In a study carried out by Miranda & Ullman (2007), 64 adults were studied using event-related potentials (ERP) which measure electrical activity in the brain. During the task, the participants listened to either familiar songs (such as *Twinkle Twinkle Little Star*) or novel songs created by the research team. For those listening to the familiar songs, there were three versions of each song: melodies containing an in-key deviant note, melodies that contained an out-of-key note, and the normal version. It was hypothesized that for the songs containing an in-key deviant note, that it would only be detected if the song was familiar and memorized; otherwise, the song would sound acceptable. However, the out-of-key note would violate rules of harmony and be noticeable even if the song was not well known. When the ERP's were analyzed, it was found that those who listened to songs that violated the memory of music corresponded to a pattern called N400. This pattern has been seen in violations with spoken word. For example, when a person is read the sentence "I'll have my coffee with sugar and concrete". Out-of-key violations of both familiar and novel melodies led to a brain-wave pattern over frontal lobe electrodes similar to patterns previously found for violations of rules in both language and music. Finally, out-of-key violations of familiar melodies also led to an N400-like pattern of brain activity, as expected because these are violations of memory as well as rules. This confirmed that music depends on two separate brain areas – memory of music being located in the temporal lobe and the general rules of music in the frontal lobe.

Understanding how music is processed can also lead to a clearer understanding of how music can affect emotions, thoughts and even the physical body. As discussed earlier, the temporal lobe is an important area in the processing of music. One structure within the temporal lobe is the amygdala, an area involved in the processing of emotional stimuli, memorizing emotional reactions and activating anxiety responses. Deficits in amygdala function have been associated with depression, anxiety and post-traumatic stress disorder among many others. This important brain area is affected when we are exposed to negative words and negative music but can be positively modulated by positive and pleasurable music (Popescu et al., 2004). Brain imaging studies have also demonstrated that music-listening generates a neural activity in the limbic and paralimbic regions which modulate both emotion and behavior (Koelsch et al., 2004, 2006; Popescu et al., 2004).

Current research has been confirming what has been known for centuries: music can both positively or negatively affect your emotions, behaviors and actions. This leads to the important discussion about how to ensure music is used to positively modulate feelings and behavior.

The Changing Brain with our use of Words

Neuroscience had traditionally accepted that the brain has localized functions and is set by adulthood. We now know that this is not true – almost every brain function is widely distributed across brain regions. In addition, there seems to be continual change based on the interactions between brain areas that are often used together. While there are critical periods of brain development, mostly in childhood and adolescents, the brain does not become stagnant after this time. The growth of new neurons and synaptic transmissions have been demonstrated to occur throughout the lifespan. These new connections are referred to as neuroplasticity.

Changes in the brain, or neuroplasticity, were also noted in the research discussed earlier on the effects of words. As we recall, with repeated exposure to negative words, various parts of the brain were negatively affected or damaged causing non-integration of its parts.

Dr. Dan Siegel has important research to contribute to our understanding of the mind-body connection. Attention can direct the flow that will activate certain neural pathways to regenerate and grow stronger. This can be summarized by Dr. Siegel's saying: *"Where attention goes, neural firing flows, and neural connection grows."* One way that you can direct attention and aid in the growth of neural connection is mindfulness. This involves the process of bringing one's attention to internal processes happening in the moment. The benefits of mindfulness include increased immunity (Davidson et al., 2003), attention enhancement (Tang et al., 2007), decreased stress (Shapiro et al., 2005) and improved cardiovascular function (Parswani et al., 2013). What you do with your mind, changes the function of your brain. A mindful focus with positive thoughts, images and words has a positive effect on the brain and can lead to positive changes throughout the physical body. Dr. Siegel also discussed how what you do with your mind, changes the function of your body, literally changing molecules that maintain the health of your cells. Integration of the brain parts is the basis of health. He discussed how non-integration of the brain creates chaos, rigidity, and disconnection, whereas harmony is created through integration of the brain parts. When we recall the research on how the corpus callosum is damaged from negative words, which connects the right and left hemispheres of the brain, it is easy to understand how non-integration of the brain occurs. This in turn affects the health of the body.

Studies have shown that when doctors and therapists teach patients to turn negative thoughts and worries into positive affirmations, the communication process improves, and the patient is able to regain confidence and self-control (Hansen & Bejenke, 2010). The use of positive words and affirmations helps activate areas of the brain involved in motivation (Alia-Kelin et al., 2007) and helps increase resilience when faced with a problem (Cohn et al., 2009). Positive words, such as "peace" or "love," can alter the expression of genes and modify areas of the frontal lobes. This area of the brain contains specific language centers that connect directly to the motor cortex, which moves the body into action. The longer you concentrate on positive words, the more you can begin to affect other areas of the brain such as the parietal lobe, which changes the perception of yourself and those you interact with. A positive view of yourself will bias you toward seeing the good in others, whereas a negative self-image will incline you toward suspicion and doubt. Over time, the thalamus will change in response to conscious words, thoughts and feelings. This can affect the way you perceive reality and respond to the world. For example, repetition of positive words like "peace" and "compassion" will activate genes that lower stress (Dusek et al., 2008), you can live longer (Fredrickson et al, 2005) and build deeper relationships with others (Losada et al., 2004).

Music and Medicine research shows how every part of the brain is stimulated with music. We hypothesize that with the use of positive lyrics in music, we are integrating the brain parts when we consider *"Where attention goes, neural firing flows, and neural connection grows."* This in turn enables the brain (control center) to manage the body (temple) at the optimal level for the soul's well-being.

Words in Improvised Music

The methodology being described in this paper will help validate this hypothesis as it has not been studied by any other research groups to date. As previously stated, much of this traditional knowledge about the power of music comes from religious ceremonies that use music to help heal the mind, body and spirit. However, scientific research in the field of spirituality and improvised music is relatively limited. The only research these writers could find relating words to improvised music was written by Alan Turry, MT called "Integrating Musical and Psychotherapeutic Thinking: Research on the Relationship Between Words and Music in Clinically Improvised Songs." This research is an interesting approach of allowing the client to creatively express his or her feelings and thoughts through improvised music. The lyrics were often negative, an expression of her inner unresolved issues.

Given that there is no known research on the use of improvised music utilizing positive lyrics in the medical field, this paper will be the first to bridge important fields together into one framework. This will aid in our understanding of the importance of using positive lyrics for its effects on the brain, which manages the body, which affects the overall well-being of the soul.

Methodology for New Holistic Approach for Healing with Improvised Music

This article will address three foundational aspects of this new holistic approach: love-based breath, positive lyrics, and improvised music. These foundational aspects are the most integral parts of the methodology for the new holistic approach to healing with music.

Love-Based Breath

If love is where all healing comes from, then it is essential that the practitioner centers him or herself in a place of love before the music healing session begins, and then maintains this meditative state during the entire process. This involves feeling and/or thinking *love* in their mind, with every breath they take during the holistic music healing session. This is very mindful work, 100% aligned with the process of where all healing comes from: love. It also creates a meditative state from which high creativity can come, allowing the process of improvisation of music and knowing when to change lyrics to become easy. When the practitioner thinks *love*, they are infusing their breath with a healing vibration, which is transmitted through the lyrics, or instrument-playing, entraining the client with a love-infused positive lyric or beautiful melody played on an instrument.

Basically, love is the underlying intention behind every breath. The client feels the love through the music, which creates a trusting relationship. The loving atmosphere creates a space where healing can begin to occur.

Positive Lyrics

The research on the effects of words on the brain, together with the Laws of Entrainment, are important in understanding the music healing session with the client – both from the practitioner's point of view, and the effect the music healing session has on the client.

While the client is engaged in the music healing session, they are under the influence of entrainment from the practitioner. In other words, they are affected by every thought and emotion of the practitioner during the music as well as the conversations. It is important that the practitioner maintain a positive attitude and use encouraging words and uplifting lyrics during the music healing session. Positive words and emotions discussed or sung are going to have a positive impact on the client's brain.

At the beginning of the session, or information gathering pre-session interview, the client will need to discuss problem areas in their life, or recall negative memories and emotions with the practitioner. From this disclosure, the practitioner can quickly assess a positive solution-based approach to alleviate the problem and sing it through the music. For example, if the client reveals fears that are prevalent that day, the practitioner goes to the opposite positive words such as "trust" or "faith", and this is what will be sung to them in the session. The practitioner will assess an appropriate positive affirmation lyric using these words depending on the context of the fear and sing it for the client. If the client is comfortable joining in, they could sing along. Alternatively, a call and response approach could be used, whereas the client sings back exactly what he/she hears from the practitioner. The client has the choice to just listen, as well. Here are some examples of positive affirmations that could be sung to the client to alleviate fears:

- *I trust that all will be okay.*
- *I trust in the universe.*
- *I have faith.*
- *I trust in my husband/children, etc.*
- *I believe that all will be okay.*

In another example, a depressed woman revealed that her mother never felt she was good enough. The practitioner could sing the following:

- You are good enough
- You are perfect, just as you are, always perfect, in every way.
- You are worthy of all good things.
- Believe in yourself, and in all you stand for.

Other positive affirmations for alleviating depression may be, but not limited to the following:

- *I forgive myself. I forgive all others for all time, for all things.*
- *I accept myself. I accept all others in their journey.*
- *I lovingly let go of the past. I have learned from the past.*
- *Peace*
- *Love*
- *I am worthy of love, of all good things.*
- *I belong*

These have been used for depression associated with grief:

- *I accept the flow of life and its cycles.*
- *I fully release everyone to his or her own journey.*
- *I lovingly let go. I accept and let go.*
- *I trust in the process of life.*
- *I have faith all will be okay.*
- *Peace*
- *Love*

During the entire session, the practitioner maintains this meditative or mindful state with these positive affirmations. Even if the client chooses not to sing along, the new positive words in the lyrics will be entrained in their brain, having positive affect for change. With repetitive practice, the client's old depressed thoughts are replaced by the new positive thoughts, which eventually become belief systems and traits. In the long term, these positive affirmations shape their personality, behavior, morals and values.

In the case where the client is a willing participant to singing along, the practitioner can carefully watch the client's reaction to each affirmation, as well as the tone in their voice, which may change as they accept these affirmations as their own. The client's tone may start out as weak, unconfident, or even full of tears at the start of the process. With continued repetition, the tone will change to a fuller, more confident, happier, and accepting lyric. It is then that the practitioner knows that the client accepts this as his or her own new truth or attitude. At that point, a new affirmation can be started. These affirmations can be assigned as homework for the client to recite or sing repetitively on their own.

Improvised Music

Using improvised music throughout the holistic music healing sessions is foundational. Then the music is specifically created for the client's unique needs in the moment. Each session's music will be different. This allows both the client and practitioner complete freedom to alter melody, rhythm, instruments, as well as lyrics in the moment. The time spent on each positive affirmation can vary session to session according to the length of time it takes for the client to accept this new higher vibration. The improvised music could use segments of known melodies or rhythms, but is open to change any element of the music at any time. The mindful state of the practitioner allows improvisation to flow freely.

Previous Research: Positive Affirmations in Pain, Anxiety and Depression

Previous research on this technique has looked at groups with anxiety, depression and pain. A study consisted of 120 participants gathered from a community sample who completed three sessions. During each one-hour session, they listened to love-based, positive affirmations put to improvised music. They were not asked to sing along in the study. In this sample, 31% had been diagnosed with a psychiatric condition at some point in their life and 65% reported some form of chronic pain in their lifetime. All healing music sessions had a random mix of persons with emotional and physical pain. A repeated measures ANOVA with a Greenhouse-Geisser correction determined that the mean reported level of depression differed significantly across sessions ($p < 0.001$). The average depression score at baseline was 10.2 out of a possible 60, whereas after the final session the average depression score was 2.5, a score noting a lack of depression. The same trend held true for anxiety ($p < 0.05$) and pain (*p*

= 0.01), decreasing mean reported pain by 4.2 points on a 10-point scale and decreasing mean anxiety 16 points on a 99-point scale.

Although the research showed improvement, the testimonials provided by the participants may be even more powerful evidence of the impact of this method. Below is a selection of testimonials from previous study sessions:

"This was profound on many levels. The pain in my right quadriceps and right thumb was released immediately. Years of physiotherapy have not had effects as amazing as these sessions."

"I had low back, nerve root damage causing sciatic pain, numbness, and spasms down left leg. The SACRED SCALE® session lessened my episodes of sciatic pain. I felt grounded and much calmer.

"I have discontinued medication for my migraines after attending the SACRED SCALE® sessions. Thank you! Thank you! Thank you!"

Common experiences ranged from tingling or rushing of cold/warm energy down a limb, to what feels like pain/tensions or stresses leaving the body. People often reported feeling much lighter afterwards, like the weight had been lifted off their hearts or off their minds. They felt more connected to their soul. They reported clarity in their thinking and increased sensitivity of their senses. Some participants even reported experiencing spiritual connections. Overall, the reports were very positive and tended to occur within 1-3 sessions, making this type of healing very fast-acting when compared to pharmacological medication. They reported improvements in mind, body and spirit.

This study shows that this form of treatment is effective in alleviating the symptoms of pain, depression and anxiety. When we looked at different forms of pain, the treatment was equally effective across pain types (i.e. migraine, joint, back, muscle pain, etc.). Also worth noting is that persons with emotional and physical pain were listening to the same positive affirmations for each session, in that we did not separate the groups into pain types. This shows that this type of therapy can help alleviate pain across a variety of symptoms and disorders. The fact that it is a passive therapy leaves it available for people suffering from a wide array of disorders that may impact language, spatial and motor functioning.

This technique has been shown to be effective across groups because it can be individualized for the problem at hand. Some examples of the positive affirmations used to alleviate anxiety/depression in the previous study are:

- *I forgive*
- *Clarity*
- *I trust in the process of life.*
- *I trust in the universe*
- *I trust all will be okay*
- *I live in the flow of life; there is time for everything.*
- *I honor my own timing in the flow of life*
- *Peace*
- *Love*

Examples of positive affirmations used for pain:

- Clarity. I have clarity. (headaches)
- I forgive my mother (father, etc.). I forgive myself. (headaches and heartaches)
- Freedom, I am free and flexible. (release of stiff tight muscles, back, neck etc.)
- I lovingly let go of the past. (holding pain in the body)
- I go with the flow of life.
- I love myself; I love all others.
- Love
- Peace

Summary

You may have noticed that the affirmations used in the holistic healing with music sessions are all Divine virtues: trust, faith, forgiveness, peace, love, letting go, etc. You may have also noted the crossover of use of these Divine virtues for physical or emotional pain. These powerful positive words, together with a love-based breath through improvised music, are the foundations of a new holistic approach to healing with music. They have the capability of transforming traumatic memories and other negative thoughts stored in the brain to a higher vibration of love, peace, faith, trust, etc. This reduces anxiety, depression, and pain in the body as evidenced by the author's study.

This new holistic approach to healing with music gives us an opportunity to consider the connection between mind, body and spirit from multi-disciplinary fields of research. It also gives us an effective way to treat anxiety, depression and physical pain. It will help us understand how mindful improvised music with love-infused positive affirmations can be used across many therapeutic settings to alleviate such imbalances. Furthermore, and perhaps most importantly, it may contribute to our own understanding of the power of our words on the wellness of our minds, bodies and souls. Future research will continue to validate the usefulness of this technique and empirically show the value that has long been understood.

References

Alia-Klein, N., Goldstein, R. Z., Tomasi, D., Zhang, L., Fagin-Jones, S., Telang, F., . . . Volkow, N. D. (2007). What is in a word? No versus Yes differentially engage the lateral orbitofrontal cortex. *Emotion, 7*(3), 649-659. doi:10.1037/1528-3542.7.3.649

Anderson C. & Carnagey (2003). Exposure to Violent Media: The Effects of Songs With Violent Lyrics on Aggressive Thoughts and Feelings. *Journal of Personality and Social Psychology,* Vol. 84, No. 5.

Aristotle ([ca. 340 BC], (1984), *The Politics.* Translated and with an introduction by Carnes Lord. Chicago: University of Chicago Press.

Barnes, Jonathan, ed. *The Complete Works of Aristotle: The Revised Oxford Translation.* Princeton: Princeton University Press, 1984. Two volumes.

Craig A. Anderson, Nicholas L. Carnagey, Janie Eubanks. (2003). Exposure to violent media: The effects of songs with violent lyrics on aggressive thoughts and feelings. *Journal of Personality and Social Psychology, 84* (5): 960 DOI: 10.1037/0022-3514.84.5.960

Cohn, M. A., Fredrickson, B. L., Brown, S. L., Mikels, J. A., & Conway, A. M. (2009). Happiness unpacked: Positive emotions increase life satisfaction by building resilience. *Emotion, 9*(3), 361-368. http://dx.doi.org/10.1037/a0015952

Davidson, R., et al. (2003). Alterations in Brain and Immune Function Produced by Mindfulness Meditation. *Psychosomatic Medicine, 65*, 564-570.

Dusek JA, Otu HH, Wohlhueter AL, Bhasin M, Zerbini LF, Joseph MG, et al. (2008) Genomic Counter-Stress Changes Induced by the Relaxation Response. PLoS ONE 3(7): e2576. https://doi.org/10.1371/journal.pone.0002576

Fredrickson, BL., & Losada MF. (2005). Positive affect and the complex dynamics of human flourishing. Am Psychol, 60(7), 678-686.

Georgetown University Medical Center. (2007, September 28). Music And Language Are Processed By The Same Brain Systems. *ScienceDaily*. Retrieved June 14, 2017 from www.sciencedaily.com/releases/2007/09/070927121101.html

Hansen, E. & Bejenke, C. (2010). Improved communication with anxious surgical patients. Anaesthesist, 59(3): 199-202.

Koelsch S, Fritz T, V Cramon DY, Müller K, Friederici AD (2006). Investigating emotion with music: An fMRI study. Hum Brain Mapp, 27: 239–50.

Koelsch S, Kasper E, Sammler D, Schulze K, Gunter T, Friederici AD (2004). Music, language and meaning: brain signatures of semantic processing. Nat Neurosci, 7: 302–7.

Losada, M. & Heaphy, E. (2004). The role of positivity and connectivity in the performance of business teams: A nonlinear dynamics model. Am Behav Scientist, 47(6): 740-765.

Miranda, R. A., & Ullman, M. T. (2007). Double dissociation between rules and memory in music: An event-related potential study. *NeuroImage*, *38*(2), 331–345. http://doi.org/10.1016/j.neuroimage.2007.07.034.

Parswani, M. J., Sharma, M. P., & Iyengar, S. (2013). Mindfulness-based stress reduction program in coronary heart disease: A randomized control trial. *International Journal of Yoga*, *6*(2), 111–117. http://doi.org/10.4103/0973-6131.113405.

Popescu M, Otsuka A, Ioannides AA (2004). Dynamics of brain activity in motor and frontal cortical areas during music listening: a magnetoencephalographic study. Neuroimage, 21, 1622–38.

Sammler, D., Baird, A., Valabreque, R., Clement, S., Dupont, S., Belin, P., & Samson, S. (2010). The relationship of lyrics and tunes in the processing of unfamiliar songs. Journal of Neuroscience, 30 (10), 3572-8. doi: 10.1523/JNEUROSCI.2751-09.2010.

Shapiro, S., et al. (2005). Mindfulness-Based Stress Reduction for Health Care Professionals: Results from a Randomized Trial. *International Journal of Stress Management, 12*(2), 164-176.

Siegel, D. (2016, September 28). A journey to the Heart of Being Human [Audio Podcast]. Retrieved from http://live.SoundsTrue.com

Talarrovicova, A., Krskove, L & Kiss, A. (2007). Some assessments of the amygdala role in suprahypothalamic neuroendocrine regulation; a minireview. Endocrine Regulations, 41 (4): 155-62.

Tang, Y., et al. (2007). Short-term meditation training improves attention and self-regulation. *PNAS, 104*(43), 17152-17156.

Tessitore A, Hariri AR, Mattay VS, Fera F, Weinberger DR. (2002) The amygdala response to emotional stimuli: a comparison of faces and scenes. Neuroimage. 17(1):317-23.

Turry, A. (2010) Integrating Musical And Psychotherapeutic Thinking: Research On The Relationship Between Words And Music In Clinically Improvised Songs. Qualitative Inquiries in Music Therapy, Volume 5, 116–172.

Mind, Memory, and Mantra

Eve Kodiak, M.M.

Abstract

The Kirtan Kriya as a Healing Tool describes the confluence of an ancient spiritual practice with medical research, citing studies that show positive effects of this activity on memory. A step-by-step guide to practicing the Kirtan Kriya is informed by speculations on why this activity can produce positive effects on the brain.

Kirtan Kriya as a Healing Tool

Memory is just one of the functions of a healthy mind/body system, but it is one of the most inconvenient ones to lose. Several studies over the past decade-and-a-half have explored the use of a particular kundalini yoga practice, the *Kirtan Kriya,* as a tool for releasing stress and improving cognitive functions. The effects are promising.

In a 2016 study[1 and 2], 14 subjects with Mild Cognitive Impairment (which can be a precursor to more serious conditions such as Alzheimer's and Dementia) showed improvement in depression, visuospatial memory, verbal memory, resilience, and executive functioning. Another study the same year[3 and 4] also showed gains in psychological well-being, mood, and sleep quality.

Researcher Dharma Singh Khalsa[5] describes some of the effects of *Kirtan Kriya*: "Practicing it may stabilize brain synapses by increasing important neurotransmitters, such as acetylcholine, norepinephrine, glutamate, and possibly GABA as well[6-8] . . . when the practitioner utilizes the fingertips in conjunction with the sound, specific areas in the brain, as seen on SPECT scans, are activated. Khalsa and Newberg showed significant cerebral blood flow changes during the practice of *Kirtan Kriya*[9] . . . the frontal lobes exhibited increased cerebral blood flow . . .the posterior cingulate gyrus (PCG) is also activated. This is relevant because the PCG is highly connected to other important anatomical areas and one of the most metabolically active regions in the brain, and it is critically important for many types of memory and emotional functions. . . additional research on KK demonstrates a diminished loss of brain volume with age[10] . . ."

Good stuff! How do we do it?

How to do the *Kirtan Kriya*

The *Kirtan Kriya* Meditation is one of the most important meditations in the Kundalini Yoga canon – according to Yogi Bhajan, who brought Kundalini Yoga to the West, it is one of three that contain within it the entire Kundalini Canon.

Kirtan Kriya is a complex activity created from several simple components that happen simultaneously. We use our voices, our fingers, and our minds to create a visual image. All three are coordinated in rhythm.

Begin by sitting in a comfortable position. If necessary, the *Kirtan Kriya* can be done lying down or standing. The important thing is that the spine is relaxed and straight. (We'll use our "calibration" time to do a few spinal flexes and perhaps some other movements to get our "instruments" ready to sit still, erect, peaceful, and relaxed for about twelve minutes).

1. **The Beat.** Rhythm precedes everything and organizes everything we do. The beat is already going, somewhere – all we need to do is catch it! The *Kirtan Kriya* uses a simple, steady four-beat pattern. Begin by just quietly hearing a "one-two-three-four" about the same speed as the second hand on a clock. You can listen to a second hand, or watch the stopwatch on your phone, or set up a metronome at 60. You can quietly tap your fingers to the beat, or lightly say some syllable – "ta, ta, ta, ta" for example. Get it into your system. This is the underlying vibration, the motor that keeps the *Kirtan Kriya* going.
2. **The Fingers**. Now that you have the beat vibrating in every cell of your body, move your fingers to it. Touch each finger to your thumb – it perfectly matches the one-two-three-four. Really feel the connection between finger- and thumb-tip. You are creating an electrical circuit; you want it to have full strength.
3. **The Mantra**. Now, instead of counting or using random syllables, we'll use four syllables to count the four-beat pattern. "Saa – Taa – Naa – Maa." "Saa" happens as the index finger connects with the thumb, "Taa" with the middle finger, "Naa" with the ring finger, and "Maa" with the baby finger.
4. **The Mouth.** "Saa Taa Naa Maa" is easy to say. Connect with the feeling of your tongue, teeth, lips and breath for each different syllable. The "S" sound is made with your tongue, teeth, and breath only. "T" and "N" require the roof of the mouth for articulation – "T" strikes higher; "N" is more forward, almost brushing against the front teeth - make sure you feel *that* connection. And finally, "M" involves the lips. This

creates a kind of micro-circular motion, beginning with a sound that opens the mouth and breath, circulating up the roof of the mouth, coming more forward, and finally closing with the lips, only to open again with the next refrain.

5. **The Tune**. Now we'll introduce the tune. Don't worry – it's a simple as the syllables. Can you sing the first line of "Mary Had a Little Lamb"? Now sing, "Mary had a" and stop. Sing those four syllables over and over. That's it - *Saa Taa Naa Maa.* For those of you who know solfege, it is equally simple. *Mi Re Do Re, Mi Re Do Re, etc.*

   It is important to keep the pitch high enough to be resounding in the chamber of the skull, rather than in the narrow pathway of the throat. The brain is higher up; we want the vibrations to reach it! And singing in the throat can create stress on the vocal cords, and stress is *not* what we are after here. It is easier to keep the sound floating when you really feel the tongue, the roof of the mouth, the teeth and the lips. It keeps the concentration in the place where the sound can resonate most easily.

6. **The Visualization.** Eyes are closed. Imagine light coming in through the top of the head (associated with the pineal gland) and making a 90 degree turn to go out the brow (associated with the pituitary gland). If you think in *chakras* – "wheels" that correspond to nerve centers in the body - the light comes in at the seventh chakra and shoots out at the sixth. If you are teaching the *Kirtan Kriya* to children, you can describe this as a really cool superhero power.

7. **The Route.** We articulate the mantra in three very different ways. First, we sing it out loud. Second, we let go of the tune and just whisper it – but with an equally clear attack and intention. Third, we articulate it silently. We make no sound, but we continue to move our mouths as if we were still whispering or singing. Then we go back to whispering, and end with singing.

8. **The Circuit.** The meridians have been described as mysterious bio-magnetic-electric pathways that keep our *qi*, or life-energy, flowing. In recent years these pathways have been studied in science as the "primo-vascular system"[11] and tracked by different

methods.[12] There are fourteen meridians. Two in particular – "mom" and "pop" – form the central power station of the body. Mom – the central meridian – runs from the perineum, up the front of the body, to the lower lip. Pop – the governing meridian – runs from the perineum, up the back of the body, over the head, and down to the upper lip. When the tongue touches the roof of the mouth, we close that circuit. As we pulse it over and over again in the *Kirtan Kriya*, that circuit is activated over and over again. This may be one of the reasons the *Kirtan Kriya* has so many positive effects.

9. **The Timing.** You can do the Kirtan Kriya for any length of time – but the instructions are to have the six parts of it (sung, whispered, silent; silent, whispered, sung) equal – or roughly equal. The studies had subjects doing each part for two minutes, which adds up to a little over twelve minutes.

How Does the Kirtan Kriya Work?

Brain imaging from the study shows that different parts of the brain became more active after doing the Kriya. It would take a lot more studies to determine why. So the following are some ideas to explore further.

1. **The Effects of Sounding.** When we hum, sing, whisper, or speak in our "head voice," we are using the skull (and jaw) as the resonant chamber for our sound. The skull is the house for the brain; the whole brain vibrates. Vibration is movement, and movement is information that translates neurologically both into reinforcing existing neural pathways, and into wiring new ones. So singing, humming, chanting, whispering, expounding, laughing – any kinds of sounds we make with energy and intention are probably helpful in growing the brain.
2. **The Effects of Rhythm and Beat.** Information organized rhythmically in our bodies and brains gets remembered. And a hypnotic beat, tune, and syllables – like those of the *Kirtan Kriya* – is front-loaded for maximum neurological effect.
3. **The Effects of Finger and Mouth Movements.** There is a universal, instinctual movement language all human beings (and animals, too) speak

unconsciously – the language of infant reflexes. Babies need to be able to survive without thinking; there are series of movement patterns that come with the rest of the developmental package to allow survival – turning, breathing, sucking, crawling, and so on. We return to these reflexive movements under stress.

One of these infant reflexes is called the *Babkin Palmomental Reflex*, and we have all seen it when we see cats kneading their paws on pillows or couches or (painfully) us. Babies do it, too, as they nurse. It helps to get the milk out. So, we all possess a neurological connection between fine motor movements and mouth movements.

When we are relaxed and integrated, we can move our mouths without moving our fingers, and vice versa. But think how many times you've seen someone writing with their tongue between their teeth. Or a child having a tantrum, screaming, while their hands work the air. Infant reflex movements return under stress, and when they are present, connections between the basal ganglia in the back of the brain and the higher functions of the brain are severed. We are wired for survival, and when we are running away from a predator, we *want* to be operating on reflexes only. To wonder whether it would be better to go left or right might mean that we become lunch.

One of the best ways to integrate infant reflexes, like the *Babkin Palmomental Reflex*, is to do these movements in a rhythmic way, in a safe environment, in an atmosphere of play and relaxation. By singing, whispering, and silently speaking while we are rhythmically moving our fingers, we are integrating the *Babkin Palmomental Reflex* and allowing the survival system to relax, creating the possibility of growing neurological connections between the primitive brain and higher order thinking.

4. **The Effects of Visualization.** When we imagine something – anything – we are activating the neural pathways[13] as if it were actually happening. Any musician or athlete knows that imagining a phrase, or a ski jump, or any activity at all, makes the execution more

effortless. As we do the *Kirtan Kriya*, we are imaginatively zapping the pre-frontal cortex (executive function, creative problem solving) roughly sixty times a minute, for twelve minutes – or 720 times. That is a lot of stimulation, especially in concert with the reinforcing finger motions and vocalizations.

Conclusion

The *Kirtan Kriya* has been used for centuries – perhaps millennia – to promote a meditative state of mind and to create a stronger relationship with the spiritual realm. It is wonderful to find that this ancient technique can still be used to help keep the brain healthy and whole.

References

1. Eyre, H. A. et al. (2016). Changes in neural connectivity and memory following a yoga intervention for older adults: a pilot study. Journal of Alzheimer's Disease, 52(2), 673–684.
2. Eyre, H. A. et al. (2016). A Randomized Controlled Trial of Kundalini Yoga in Mild Cognitive Impairment
3. Kim E. Innes, et al (2016) Effects of Meditation versus Music Listening on Perceived Stress, Mood, Sleep, and Quality of Life in Adults with Early Memory Loss: A Pilot Randomized Controlled Trial. Journal of Alzheimer's Disease 52 (2016) 1277–1298 DOI 10.3233/JAD-151106 IOS Press
4. Kim E. Innes, et al (2016) Meditation and Music Improve Memory and Cognitive Function in Adults with Subjective Cognitive Decline: A Pilot Randomized Controlled Trial. Journal of Alzheimer's Disease 56 (2017) 899–916 DOI 10.3233/JAD-160867 IOS Press
5. Dharma Singh Khalsa (2015), Stress, Meditation, and Alzheimer's Disease Prevention: Where the Evidence Stands, Journal of Alzheimer's Disease, vol. 48, no. 1, pp. 1-12, 2015 http://content.iospress.com/articles/journal-of-alzheimers-disease/jad142766
6. Newberg AB, Iversen J (2003) The neural basis of the complex mental task of meditation: Neurotransmitter and neurochemical considerations Med Hypotheses 61: 282291
7. Streeter CC, Jensen JE, Perlmutter RM, Cabral HJ, Tian H, Terhune DB (2007) Yoga asana sessions increase brain GABA levels: A pilot study J Altern Compl Med 13: 419426

8. Nash JD, Newberg A (2013) Towards a unifying taxonomy and definition for meditation Front Psychol 4: 806
9. Khalsa DS, Amen D, Hanks C, Money N, Newberg AN (2009) Cerebral blood flow changes during chanting meditation Nucl Med Commun 30: 16
10. Newberg AB, Wintering N, Khalsa DS, Roggenkamp H, Waldman RW (2010) Meditation effects on cognitive function and cerebral blood flow in subjects with memory loss: A preliminary study Alzheimer's Disease 20: 517526
11. **Miroslav Stefanov The Primo Vascular System as a New Anatomical System, Journal of Acupuncture and Meridian Studies, Volume 6, Issue 6, December 2013, Pages 331-33** http://ac.els-cdn.com/S2005290113002082/1-s2.0-S2005290113002082-main.pdf?_tid=e9c8a69e-9a0c-11e7-bf30-00000aab0f27&acdnat=1505476771_ab46157abd55059ef72b3e0e293f94c2
12. C. Will Chen (2013) Wave-Induced Flow in Meridians Demonstrated Using Photoluminescent Bioceramic Material on Acupuncture Points Evidence-Based Complementary and Alternative Medicine 2013; 2013: 739293. Published online 2013 Nov
7. doi: 10.1155/2013/739293
13. Istvan Molnar-Szakacs, Music and Mirror Neurons, from Motion to "e" motion, Social, Cognitive and Affective Neuroscience, Dec. 2013 https://www.ncbi.nlm.nih.gov/pmc/articles/PMC3838801/

Numerology in Classical Chinese Music and Medicine: An Exploration of Mutual Significance and Potential Clinical Applications

Dena Gold

Abstract

The ancient Chinese worldview contains a complex series of interrelated ideas that weave their way throughout a diverse array of sciences and arts; ancient Chinese numerology is one such idea. While not often explicitly discussed, it manifests in many fields including music and medicine. The specific numbers that are most prominent between these two disciplines are five (phases/elements), six (joists/harmonies), ten (celestial stems), and twelve (terrestrial branches). There is an implicit interaction here between heaven, earth and man. The 5/6 differential being that of the inner and outer world and the 10/12 being the differential between heaven and earth. This paper will seek to explore some of the ways these numbers appear in ancient Chinese music theory and will look to further understand the relationships of the 12 rules, 5 notes, etc. It will then seek to propose potential clinical applications based upon the correlative systems. As bell therapy was used to treat the body-country of the emperor, we know there is some tradition of sound therapy in Chinese medical history, while we certainly would require more inquiry to discover how exactly that was used, knowing there were sixty bells (5x12) does offer some space for speculation, which will be explored in later sections.

Music

In most ancient cultures there is a deep spiritual appreciation of sound and music. In the Sufi tradition of Hazrat Inayat Khan, the cascade of manifestation of existence is explained as originating from Source, first manifesting as sound, and then as light, making that which can be heard one degree closer to the Source than that which can be seen (Khan 1996, 144-5). The Raga system[1] of India appreciates that certain notes and keys can generate different moods and states, and thus are more appropriately played at certain times of day and in certain contexts. Hamza al-Din of Sudan translated Avicenna's work on music (though died before completing this translation), discussing the effect of certain notes on a human being. The Chinese also have such an appreciation of music, though, just as all of these are somewhat different based on their culture's music systems, the ancient Chinese primarily used their own indigenous understanding of the construction of the universe and the significance of certain numbers to appreciate what sound is, what it can do, and how to work with it at the level of an individual.

In his article "The Sound and the Fury" (2014), Stephen Jackowicz describes the ancient Chinese view of their language as Adamic[2], meaning that the words of the language were an exact expression of, and carry some force of what was being described, rather than being solely a tool of description. Sound itself was not just performative, but interactive. Furthermore, the Chinese character for music (樂 *yue*) is actually the same as the word used to describe herbal medicine (藥 *yao*), minus the radical for "grass". We can make a conceptual leap and presume the application of herbal medicine and of music were in some way the same – external forces applied to the body to help rectify imbalance.

Numerological significance

In many cultures, there is a tendency to ascribe significance to certain numbers, often because they relate to religious

[1] Raga is somewhat akin to Western musical modes, though has more nuances, such as certain notes that are played differently on ascent versus descent, certain trills, etc.

[2] Other ancient cultures had this belief about a number of languages (Hebrew, Sanksrit, Arabic, etc.).

symbolism. One simple example of this is the absence of a 13th floor in the United States because the number 13 is considered unlucky. This is a symbolism adopted from Christianity, as Judas was the 13th guest to sit at the Last Supper and is the apostle who betrayed Jesus. In Judaism, there is an entire song sung at the Passover Seder going through numerological correlations, and there is a system of assigning number values to words (גמטריה *gematriyah*), which is why the number 18, the numerological value of the word "חי" (*hai,* life), is considered lucky. In ancient China, there is also such a system of correlation called appearance of numbers (象數 *xiangshu*), which is delineated in Table 1 below:

Table 1. Showing Chinese systems of correlation called appearance of numbers (象數 *xiangshu*)

Number	Principal appearance	Common manifestation
0	太虛 the great emptiness	道 the *Dao*
1	太極 the *taiji*	This is the "great limit," it is the precursor to division, a state of singularity. In some interpretations it is thought of as the source of all things.
2	兩儀 two rites	陰陽 *yin* and *yang*
3	三才 three talents	These are 天人地 heaven, man, earth. Also can be seen as the 三家 Three houses (of time): past, present, future
4	四象 four appearances	太陽，少陽，太陰，少陰 greater yang, lesser yang, greater yin, lesser yin
5	五行 five phases	火土金水木 fire, earth, metal, water, wood
6	六合 six joists/harmonies	東南西北天地 east, north, west, south, up, down
7	七星 seven stars (of the big dipper)	These are: 天樞，天璇，天璣，天權，玉衡，開陽，瑤光 , stars Celestial Pivot/Dubhe, Celestial Rotating Jade/Merak, Celestial Shining Pearl/Phecda, Celestial Balance/Megraz, Jade Sighting-Tube/Alioth, Opener of Yang/Mizar, Twinkling Brilliance/Alkaid. In a human being we tend to think of the number seven as correlating with the emotions: 怒 motivation/anger, 喜 joy/mania, 憂 planning/worry, 悲 compassion/sadness, 恐 caution/fear, 驚 fright, 思 overthinking.
8	八卦 eight trigrams	The eight *gua* are: 乾, 坤, 震, 巽, 坎, 離, 艮, 兌 *qian*/heaven, *kun*/earth, *zhen*/thunder, *xun*/wind, *kan*/water, *li*/fire, *gen*/mountain, *dui*/lake Also thought of as the four directions plus their intermediary directions (N, NW, W, SW, S, SE, E, NE)

Table 1 (cont'd). Showing Chinese systems of correlation called appearance of numbers (象數 *xiangshu*)

Number	Principal appearance	Common manifestation
9	九宮 nine palaces	This is the arrangement of the eight *gua* in a square, with a central point. Each one is represented by a number, with five in the middle and each line adding up to fifteen, known as the *luo shu* (洛書). In a human being we correlate them with the 九竅 nine orifices.
10	十天根 ten celestial stems	甲，乙，丙，丁，戊，己，庚，辛，壬，癸 *jia*/*yang*-wood, *yi*/*yin*-wood, *bing*/*yang*-fire, *ding*/yin-fire, *wu*/*yang*-earth, *ji*/*yin*-earth, *geng*/*yang*-metal, *xin*/*yin*-metal, *ren*/*yang*-water, *gui*/*yin*-water.
11	十一度 eleventh degree	This is the interaction between the stems and branches. This degree is the differential that keeps dynamism in the interaction between the two
12	十二地支 twelve terrestrial branches	子，丑，寅，卯，辰，巳，午，未，申，酉，戌，亥 *zi*/rat, *chou*/ox, *yin*/tiger, *mao*/rabbit, *chen*/dragon, *si*/snake, *wu*/horse, *wei*/goat, *shen*/monkey, *you*/rooster, *xu*/dog, *hai*/pig

Numbers in Music

Therefore the wise ruler circulates through the twelve provinces, guarding each of their customs and manners, reviews each of their natures and emotions, thusly he deduces the twelve customs, sets the six rules (as in pitch-pipes), five sounds, eight tones, and seven beginnings.

故聖王巡十有二州，觀其風俗，習其性情，因論十有二俗，定以六律，五聲，八音七始 (*Shangshu da zhuan* quote taken from Brindley 2012, 58)

We can see from the above quote that the musical notes themselves were established via the emperor's observation of the realm, as music was a way of maintaining harmony in the kingdom (for more information on this, see Brindley 2012). This also gives us one system of correlation between numbers and music. Brindley, while acknowledging that we don't have enough exact information about ancient music practices to give a highly precise interpretation of this passage, does give a sufficient explanation of each of these numerological significances: twelve as the terrestrial branches and twelve pitch-pipes, derived from the six mentioned in the quote[3], the five tones as correlating to the four directions plus the center, the seven beginnings as the seven specific pitch standard correlating to the four directions, the three talents (heaven, earth man), and eight relating to the different materials of instruments[4] (Brindley 2012, 59).

Ancient Chinese music, often referred to as the *GongDiao* system, is actually two different systems combined, *GongYin* (宫音) and *Diao*[5]*Sheng* (調聲). Chen Ying-shi very clearly delineates how the notes of the *GongYin* system can move around the *DiaoSheng* system, as they are actually two different systems (Chen 2002, 110-111). The *gong* notes are not static, but highly mutable and based on relationship to one another rather than hertz value. This idea is important to clarify before moving into the more abstract concepts of these two. The *Gong* system is pentatonic, the *Diao* system (that is the system that uses the *huangzhong,* or yellow bell, as its tonic) is a twelve-note scale, similar to the chromatic scale of Western music.

The *Gong* system is composed of five scale degrees: *gong* (宫), *shang* (商), *jue* (角), *zhi* (徵), and *yu* (羽). The number five is highly prominent in East Asian medicine, as it correlates with the five phases (otherwise referred to as the "five elements"), which have a correlation to the organs, and thus the channels of acupuncture. Often these are considered to be static notes (C, D, E, G, and A, respectively), but more accurately, they can be thought of as intervals. The *gong* note is the first tone in the scale, and is associated with the earth phase. *Shang* is the second scale degree, a whole step down from *gong*. *Jue* is associated with wood and is the third note in the scale, two steps down from the tonic. *Zhi* is associated with fire and is three-and-a-half steps from the tonic. *Yu* is associated with the water phase and is four-and-a-half steps from the tonic.

It is worth noting that the way that pitches were arrived at in early Chinese music theory, as per Hafez Modirzadeh (2001, 81), is via a system of falling fourths

[3] While she does not explain these six tones from which the twelve are derived, it is possible to think of them as the twelve rising and falling pairs, as will be explained in the following paragraphs.

[4] It is possible that this aspect of music could be used to treat the eight extraordinary vessels, but that is far beyond the scope of this paper.

[5] Chen also points out this could be alternatively pronounced *tiao* (harmonize/blend/adjust/regulate).

and rising fifths, which he demonstrates in the Figure 1 below:

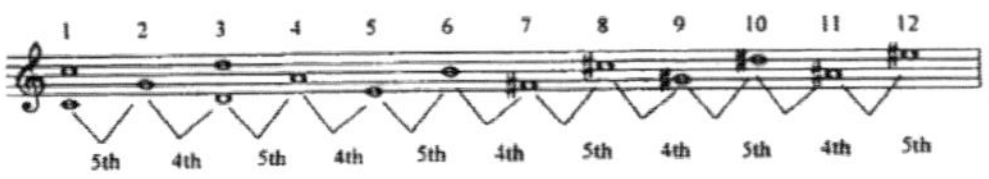

Figure 1. Showing the falling fourths and rising fifths

Differentials in Chinese Philosophy

This is an interesting phenomenon in the Chinese understanding that does not force things to fit neatly into the same numerological box, but actually supports differentials as a force of dynamism. For example, in regard to humans, the body is said to be organized around five, while the external world is said to be organized around six. This differential between the human body and the external environment is the reason why one ages. In terms of measures of time, the ten celestial stems and twelve earthly branches are two systems that are overlaid to measure days, months, and years. The differential here (between the ten celestial stems and the twelve terrestrial branches) is what allows time to move forward. If the two were perfectly matched, time would stop at the end of the first cycle (this is the significance of the eleventh degree). In music, despite the fact that a chromatic scale has an even number of notes (twelve), and could be divided evenly by six, the scales that are most significant are made of five and seven notes (which do coincidentally, or perhaps not so coincidentally, add to twelve). Furthermore, the perfect intervals also are five half-steps and seven half-steps from the tonic, while the tritone (that is the one halfway through the chromatic scale, six half steps from the tonic) was rumored to have been banned by the Catholic Church, or at least popularly shunned by church musicians for a period of time and called "*diabolus in musica*" (the devil in music) because of its dissonant nature. The use of the perfect fourth and perfect fifth in determining the twelve-note scale speaks to this differential (as opposed to the methods used for even-tempered scales). The perception of this is seen as an asymmetrical spiral rather than a congruent circle, only coming to a complete octave relationship to the root at the fifty-third operation (Modirzadeh 2001, 85). It is worth noting that when it is taken to this degree, it is often taken to sixty notes (interesting - as this is five multiplied by twelve).

Two

In this system, the first division that we see is that of *yin* (陰) and *yang* (陽), of rising fifths and falling fourths. The *yang* tones are the descending fourth (C, D, E, G♭, A♭, B♭), and the *yin* tones are the ascending fifth (C♯, D♯, F, G, A, B) (Modirzadeh 2001, 82, via Danielou 1995, 34-35, 41). *Yin* and *yang* are the first division. The *Dao* (道) is considered the One, and the first division is the bifurcation of *yin* and *yang*. These two principles are considered relative to one another. *Yang* is more masculine, dispersing, lighter, brighter, celestial, and dynamic; *Yin* is more feminine, condensing, heavier, darker, terrestrial, and static. These two are intimately intertwined in a relationship of relativity.

The implications of *yin* and *yang* in East Asian medicine are far-reaching. Most everything is subject to this binary grouping, with the understanding that there is some level of construction, as this relationship is inherently relative. Firstly, *yin* and *yang* are considered actual textures within the body. The organs and channels are grouped into *yin* and *yang* categories as well. There is little in the body that does not follow this grouping, as it is the primary division in Chinese thought.

Use of purely *yin* and *yang* is considered the most complex and refined system of diagnosis in East Asian medicine, as the division into two is the closest to wholeness, containing only one single division. It is hardest to cultivate the ability to work within this system, and thus it is also hard to prescribe how this could be equally applied to music. This pairing is omnipresent, as all things have a relative relationship of *yin* and *yang*, and thus is integral to understanding most other concepts within the ancient Chinese concept.

Five

The Chinese character 五 (*wu,* five) according to Elisabeth Rochat, "represents the crossing in motion, between *yin* and *yang* or heaven and earth, or two poles of vitality. You have the construction or appearance of a centre by the crossing of different qualities of *qi*, and the ability to build something from that," (Larre and Rochat 1996, 23). Therefore, here we see the number five creates some relationship of motion from the center, or more accurately, creates a center via the dynamic interaction of forces. In Chinese thought, the center is generally placed within the division of the *wuxing* (五行), the five phases. We can understand this relationship with Larre and Rochat's explanation, as there is a diagram of the five with earth in the middle of the two *yin-yang* axes of metal-wood and fire-water.

These five phases are intimately interrelated to each other, much like *yin* and *yang*, but in multiple dynamics, rather than just the binary relationship of the two. They are superimposed in two major cycles, that of generation (生 *sheng*) and that of control (克 *ke*). These two major cycles of interrelationships and interaction can be drawn as a circle inscribed with a five-pointed star or pentagram. The sequence of the cycle of generation follows along the circle's perimeter in a clockwise direction; wood generates fire, fire generates earth, earth generates metal, metal generates water, and water generates wood. The sequence of the controlling cycle skips one phase, inscribing a five-pointed star or pentagram on the inside of this circle. Wood controls earth, earth controls water, water controls fire, fire controls metal, and metal controls wood. These interrelationships maintain balance between the five phases[6]. Figure 2 shows the two variant models of the five phases side by side:

Figure 2. showing the two variant models of the five phases side by side

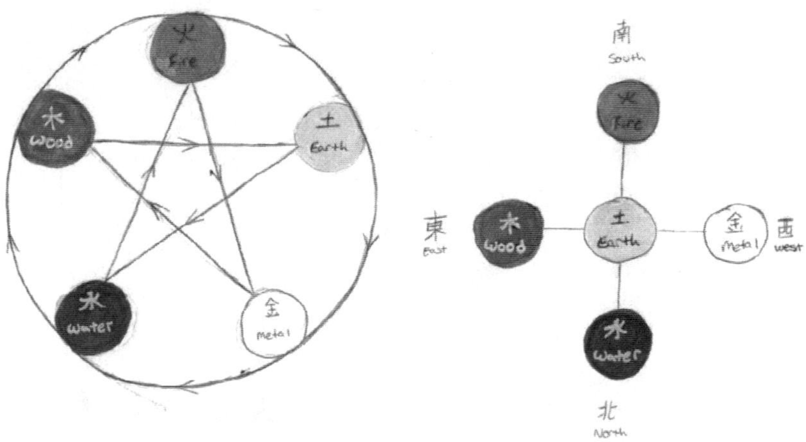

[6] For more information on the five phase interactions, see *The Web that has no Weaver*, by Ted Kaptchuk 437-450 (Contemporary Books, 2000).

Table 2. Showing the five phase correlations:

Phase	Fire	Earth	Metal	Water	Wood
Color	Red	Yellow	White	Black	Green
Sound	Laughing	Singing	Weeping	Groaning	Shouting
Odor	Scorched	Fragrant	Rotten	Putrid	Rancid
Flavor	Bitter	Sweet	Pungent	Salty	Sour
Solid Organ	Heart/ pericardium	Spleen	Lung	Kidney	Liver
Hollow organ	Small intestine/ triple heater	Stomach	Large intestine	Urinary bladder	Gallbladder
Emotion	Joy/mania	Planning/worry	Grief/sadness	Caution/fear	Motivation/anger
Spirit	Spirit 神	Intention 意	Corporeal soul 魄	Will 志	Ethereal soul 魂
Note	*Zhi* 徵	*Gong* 宫	*Shang* 商	*Yu* 羽	*Jue* 角

In the body there are five spirits (or really categories of spirits[7]), which correlate to the associated paired solid/yin organ[8] (藏 *zang*). These are the spirit (神 *shen*) related to the heart (心 *xin*), the intent/reasoning function (意 *yi*) related to the spleen (脾 *pi*), the corporeal soul(s) (魄 *po*) related to the lung (肺 *fei*), the will (志 *zhi*) related to the kidney (腎 *shen*), and the ethereal soul(s) (魂 *hun*) related to the liver (肝 *gan*).

The division of notes into these categories follows a clear schema of intervals outlined above, and is often overlaid on the *DiaoSheng* system by using an inner and outer wheel to determine the notes (Chen 2002, 112):

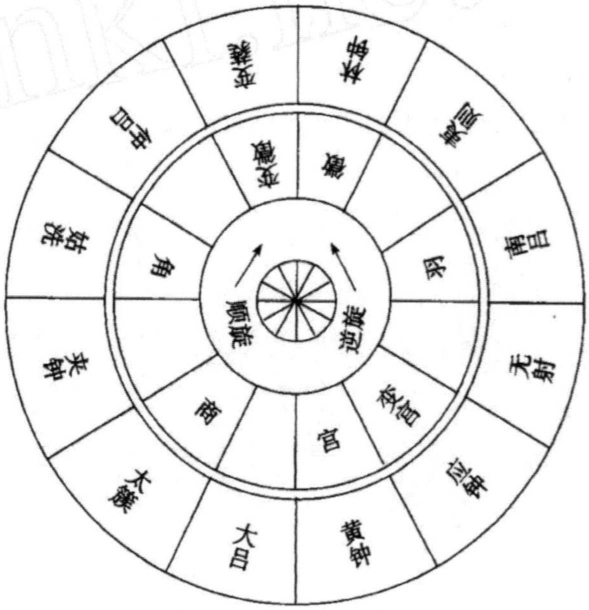

Figure 3. Showing the five phase scale plus the two *bian* notes overlaid on the twelve notes of the *huangzhong* system. Note the inner wheel can be turned to start on any of the twelve.

[7] Of the five spirits, two are considered to have multiple spirits of the same category within one body: the *po* (魄) are numbered as seven (or twenty-seven) and the *hun* (魂) as three.

[8] The organs of the body are divided into *yin*, solid organs and *yang*, hollow organs.

There is more in this image that can be discussed, as the inner wheel does also have the extra two *bian* notes making a scale of seven, which will be discussed later, but the primary value for the purposes of this discussion in this section is the relationship between five and twelve. This wheel shows how the *gong* scale can move around the *diao* scale.

Using C as the tonic, the order of generation of the scale via the ascending fifth/descending fourth would be: C, G, D, A, E – earth, fire, metal, water, wood. Those are actually the first five notes generated via this rising-fifth-falling-fourth process, and the two additional *bian* notes are the next two generated. This order seems better suited to the former representation of the five phases based on the four cardinal directions, with earth drawn in the middle, with fire south, water north, metal west, and wood east. The line would be drawn from the center down and then follow the four cardinal points in a clockwise direction. This would also allow the arrangement of the notes in ascending order to move from center to the east-west axis, and then the north-south axis. Each of the twelve main meridians have five points representing each phase on them, making a total of sixty five-phase points on the body. A musician then could accomplish the same task as an acupuncturist with a needle with the appropriate understanding of notes in relationship to phases. Protocols that work within the acupuncture system could be transposed onto a musical one. For example, the Korean Sa'am five element acupuncture system[9] is one of dealing with excesses and deficiencies via mother/child relationships - that is the outside circle of the pentagram, going clockwise, or the generating cycle. For example, to tonify the heart, one tonifies Ht 9 and Lv 1, and disperses Ht 3 and Kd 10. This is tonifying the mother point (e.g. the wood point) on the "self," or affected channel and on the mother channel, and dispersing the controlling point (e.g. the water point) on the "self", or affected channel and the controlling channel. It is worth noting that this second set of points can change depending on the pattern presentation. The following chart summarizes this technique.

[9] For more information see Willmont 2010, 228-243. This is just one system that uses the five phase point correlations to determine treatment, but any other would also be appropriate to apply.

Table 3. Showing the points for the Korean Sa'am five element acupuncture system

Channel	To Tonify					To Disperse			
	Tonify		Disperse			Tonify		Disperse	
Lung	Lu9	Sp3	Lu10	Ht8		Lu10	Ht8	Lu5	Kid10
Large Intestine	Ll11	St36	Ll5	Sl5		Ll5	Sl5	Ll2	UB66
Stomach	St41	Sl5	St43	Gb41		St43	Gb41	St45	Ll1
Spleen	Sp2	Ht8	Sp1	Lv1		Sp1	Lv1	Sp5	Lu8
Heart	Ht9	Lv1	Ht3	Kd10		Ht3	Kd10	Ht7	Sp3
Small Intestine	Sl3	Gb41	Sl2	UB66		Sl2	UB66	Sl8	St36
Urinary Bladder	UB67	Ll1	UB40	St36		UB40	St36	UB65	Gb41
Kidney	Kd7	Lu8	Kd3	Sp3		Kd3	Sp3	Kd1	Lv1
Pericardium	Pc9	Lv1	Pc3	Kd10		Pc3	Kd10	Pc7	Sp3
Triple Heater	TH3	Gb41	TH2	UB66		TH2	UB66	TH10	St36
Gallbladder	Gb43	UB66	Gb44	Ll1		Gb44	Ll1	Gb38	Sl5
Liver	Lv8	Kd10	Lv4	Lu8		Lv4	Lu8	Lv2	Ht8

The color-coding is with the colors traditionally associated with each element: red for fire, yellow for earth, white for metal, black for water, and green for wood.

Six

The number six, while not of primary importance to the topic at hand, should be briefly discussed, as twelve is six doubled. Six is correlated with the six joists/harmonies (六合 *liu he*), which is the interface with the outside, three-dimensional world; they are: front, behind, left, right, above, and below.

Rochat mentions that while "五" (*wu,* five) is a place with crossings of different qualities and a place from which things can be built, "六" (*liu,* six) is life within, constructing itself. It is like an inner maintenance, with all the interplay of *yin* and *yang* allowing my own life to develop and maintain itself. Therefore you have six meridians or six *fu*," (Larre and Rochat 1996, 23). And in the *ZuoZhuan*, "Heaven has six *qi*, in descent they generate the five flavors, in emission they become the five colors, in diminishment they become the five sounds (天有六氣，降生五味，發為五色，微為五聲)," and so, the phenomena of heaven[10] relating to the external world[11]

[10] In this context the "six *qi* of heaven" means the weather, i.e., the six climatic factors: heat, cold, dampness, dryness, wind, and summerheat.

[11] This is in juxtaposition to the internal environment, as the East Asian medical model views the internal workings of the body as comparable to those of the external, natural world (i.e. comparing meridians to rivers, *wei qi* as mist, etc.)

can be seen to align with the number six. Thus, we see that juxtaposition of the five solid and the six hollow organs. The number five, relating to the *yin* organs, is more structural and are the building blocks of potential, while the number six, relating to the *yang* organs, is more dynamic and is the place of interaction between these, and thus the place of construction. While the five phase system relates to the inner working of the body, the six joists are the meeting with the external, which is why the number six in east Asian medicine can related to the six hollow organs (all of which have contact with the outside) and the six factors which cause disease (when aspects from the outside that do not belong inside the body enter it).

It is also interesting to note that in both music and medicine in the ancient Chinese model, when working within the system of twelve, it is not "六" (*liu,* six)[12] that is in the foreground, but "五" (*wu,* five) and "七" (*qi,* seven), keeping in line with the idea of differential models which are the source of dynamism.

Seven

According to Rochat, "The character for seven (*qi* 七) is drawn with a horizontal line, the earth, and from underground something is pushing itself up and springing forth. It is the only one of the number which has real aspirations in its pronunciation," and that "[seven is] the piercing of passages between the exterior and the interior" (Larre and Rochat 1996, 23-24).

Seven is the number of stars in the big dipper, which is considered the right wall of the Purple Forbidden Enclosure, connecting back to this idea of passages between exterior and interior. Seven is the number of notes in the pentatonic scale plus the *biangong* (變宮) and *bianzhi* (變徵) notes, which makes it similar to typical Western scales.

In East Asian medicine, the seven emotions (七情 *qi qing*) are the seven internal causes of disease. There is more than one model, but the typical medical one is: joy/mania (喜 *xi*), anger/motivation (怒 *nu*), worry/planning (憂 *you*), sadness/grief

[12] However, the western music traditions do have the hexachord system, which in an interdisciplinary model could be used to work with forces in this system of six.

(悲 *bei*), fear/caution (恐 *kong*), fright (驚 *jing*), and over-thinking (思 *si*)[13]. Joy is associated with the fire phase, anger with wood, worry with earth, sadness with lung, and fear with water. The remaining two, fright and over-thinking, lay outside of this model, as each of the others have both a positive and negative aspect, while the latter two of these are only pathological. To accord these with the scale, it would make the most sense to link the five emotions, which correspond with the five phases, with their respective notes, while correlating the two *bian* notes with these latter two emotions. The *biangong*, being a variation on *gong*, would most clearly relate to over-thinking, as it is related to general cognitive function, which is analogous to the spirit of the spleen (意 *yi*). The last note, the *bianzhi*, would be associated with fright, most clearly related via the connection to the blood[14].

The connection in Chinese thought between music and the emotion is well established. There are differing viewpoints that argue for music having the ability to cause certain emotions to arise in someone versus those that argue that it just arouses what is already latent within the individual (Middendorf 2010, 137-8)

The damaging effects of the five phase emotions plus over-thinking (思 *si*) are listed in the *LingShu* (靈樞) chapter eight. There it says that apprehension and over-thinking injure the spirit (心，怵惕思慮則傷神). Worry and anxiety non-stop injure the *yi* (脾愁憂而不解則傷意). Grieving injures the *hun* (肝悲哀動中則傷魂). Mania injures the *po* (肺喜樂無極則傷魄). Anger injures the will (腎盛怒而不止則傷志). Fear without resolution injures the essence (恐懼而不解則傷精). And the five phase emotions' physiological affects and correlative movements are listed in *SuWen* (素聞) chapter thirty nine:

[13] Alternately, there are systems that view repression (鬱 *yu*) as the seventh emotion rather than over-thinking. In that model, the *biangong*, falling between the *yu* and *gong* note, would be most logically correlated with fright, and the *bianzhi* falling between *jue* and *zhi* would most logically correlate with repression. Fright can be related to the kidney and over-thinking relating to spleen.

[14] Fright is thought to momentarily extinguish *mingmen* fire, causing a nugget of blood to freeze and remain frozen.

Anger: then *qi* rebels, this then [causes] vomiting of blood and food to leak out, therefore *qi* is ascending. Happiness: then the *qi* is harmonized and reaches the will, the *ying* and *wei* communicate favorably, therefore the *qi* is leisurely. Sadness: then the heart system is urgent, the lung spreads, the lobes lift and the upper burner does not course, the *ying* and *wei* are not disseminated, [there is] oppressive heat in the middle, and therefore *qi* disappears. Fear: then the essence declines, then the upper burner is obstructed, [and due to this] obstruction the *qi* then returns, returning then the lower *jiao* is swollen, therefore *qi* does not move. Cold: then the interstices are obstructed, the *qi* does not move, therefore the *qi* is restrained. Heat: then the interstices are opened, the *ying* and *wei* course through the sweat, there is great draining, therefore the *qi* drains. Fright: then the heart is without a place to lean, the spirit is without a place to return, consideration is without a place to settle, therefore the *qi* is disordered. Toil: then gasping for breath and sweat emits, outside and inside are both excessive, therefore the *qi* is wasted. Thinking: then the heart has a place of manifestation, the spirit has a place to return, the upright *qi* remains and does not move, therefore *qi* is knotted.

怒則氣逆，甚則嘔血及飧泄，故氣上矣。喜則氣和志達，榮衛通利，故氣緩矣。悲則心系急，肺布葉舉，而上焦不通，榮衛不散，熱氣在中，故氣消矣。恐則精卻，卻則上焦閉，閉則氣還，還則下焦脹，故氣不行矣。寒則腠理閉，氣不行，故氣收矣。炅則腠理開，榮衛通汗大泄，故氣泄。驚則心無所倚，神無所歸，慮無所定，故氣亂矣。勞則喘息汗出，外內皆越，故氣耗矣。思則心有所存，神有所歸，正氣留而不行，故氣結矣。

(*SuWen* 39:6)

Combining these two can generate Table 4 below:

In other schema, over-thinking is the primary emotion of the spleen, while worry/anxiety becomes associated with the lung. If that were to be the paradigm used, then the associated note would change accordingly.

Ten

The number ten is aligned with the number of celestial stems, which in essence is the breaking down of the five phases into yin and yang pairs. They are: *jia* (甲) *yang*-wood, *yi* (乙) *yin*-wood, *bing* (丙) *yang*-fire, *ding* (丁) *yin*-fire, *wu* (戊) *yang*-earth, *ji* (己) *yin*-earth, *geng* (庚) *yang*-metal, *xin* (辛) *yin*-metal, *ren* (壬) *yang*-water, *gui* (癸) *yin*-water. While this is not a prevalent number in either music or East Asian medicine on its own, it is worth noting as the doubling of five, by dividing five into *yin* and *yang* pairs. As the celestial counterpart to the terrestrial twelve, there could be scales built upon a base of ten, and potential treatment methods based on this. This would require its own research and reasoning, and is currently beyond the scope of this paper.

Table 4. Showing the seven emotions, their associated organ, movement, and musical note

Emotion 情 *qing*	Organ 藏 *zang*	Movement 行 *xing*	Injures 傷 *shang*	Associated note
Anger 怒 *nu*	Liver 肝 *gan*	Ascent 上 *shang*	Will 志 *zhi*	*Jue* 角
Pleasure/mania 喜 *xi*	Heart 心 *xin*	Leisure 緩 *huan*	Corporeal soul 魄 *po*	*Zhi* 徵
Anxiety/worry 憂 *you*	Spleen 脾 *pi*		Intent 意 *yi*	*Gong* 宮
Sadness 悲 *bei*	Lung 肺 *fei*	Disappear 消 *xiao*	Ethereal soul 魂 *hun*	*Shang* 商
Fear 恐 *kong*	Kidney 腎 *shen*	Does not move 不行 *buxing*	Essence 精 *jing*	*Yu* 羽
Fright 驚 *jing*		Disordered 亂 *luan*		*BianZhi* 變徵
Over-thinking 思 *si*		Knotted 結 *jie*	Spirit 神 *shen*	*BianGong* 變宮

Twelve

Twelve is the number of terrestrial branches (地支). These along with the celestial stems are the measure of time in terms of hour, day, month, and year (Brindley 2012, 59). The overlaying of the system of ten on top of twelve gives us sixty combinations, as they move sequentially and are both even numbers, allowing only even to pair with even and odd to pair with odd – in this case meaning only certain stems can pair with *yin* branches and others with *yang* branches. The terrestrial branches are most commonly known as the twelve animals of the zodiac; this is certainly a cursory explanation. They are cycles of time in both small and large scale; they are unseen forces that set a certain tone for the timeframe they govern. In the modern world, this is the most common division of the octave, and is the number arrived at when one uses only the perfect fifth (and/or its inverse, the perfect fourth) to generate the notes of the octave.

The *DiaoSheng* system begins at *huangzhong* (黃鐘), the tonic, the progresses through to the *dalü* (大呂, the semitone), *taicu* (太簇, the major second), *jiazhong* (夾鐘, the minor third), *guxian* (姑洗, the major third), *zhonglü* (仲呂, the perfect fourth), *ruibin* (蕤賓, the tritone/wolf interval), *linzhong* (林鐘, the perfect fifth), *yize* (夷則, the minor sixth), *nanlü* (南呂, the major sixth), *wuyi* (無射, the minor seventh), *yingzhong* (應鐘, the major seventh).

The terrestrial branches are as follows: *zi* (子, 11 PM – 1 AM/11th month), *chou* (丑, 1 AM – 3 AM/12th month), *yin* (寅, 3 AM – 5 AM/1st month), *mao* (卯, 5 AM – 7 AM/2nd month), *chen* (辰, 7 AM – 9 AM/3rd month), *si* (巳, 9 AM – 11 AM/4th month), *wu* (午, 11 AM – 1 PM/5th month), *wei* (未, 1 PM – 3 PM/6th month), *shen* (申, 3 PM – 5 PM/7th month), *you* (酉, 5 PM – 7 PM/8th month), *xu* (戌, 7 PM – 9 PM/9th month), and *hai* (亥, 9 PM – 11 PM/10th month). Each one of these correlates with an acupuncture channel, the one in which the *qi* predominates during the corresponding time. Table 5 summarizes the branches with their correlated scale degrees, meridians, and organs:

Table 5. Showing summary of the branches with their correlated scale degrees, meridians, and organs.

Musical Pitch 律呂		Branch 地支	Meridian 經絡	Organ 臟腑
HuangZhong 黃鐘	1	*Zi* 子	Gallbladder 膽	Kidney 腎
DaLü 大呂	2	*Chou* 丑	Liver 肝	Spleen 脾
TaiCu 太簇	3	*Yin* 寅	Lung 肺	Gallbladder 膽
JiaZhong 夾鐘	4	*Mao* 卯	Large Intestine 大腸	Liver 肝
GuXian 姑洗	5	*Chen* 辰	Stomach 胃	Stomach 胃
ZhongLü 仲呂	6	*Si* 巳	Spleen 脾	Small Intestine 小腸
RuiBin 蕤賓	7	*Wu* 午	Heart 心	Heart 心
LinZhong 林鐘	8	*Wei* 未	Small Intestine 小腸	Spleen 脾
YiZe 夷則	9	*Shen* 申	Urinary Bladder 膀胱	Large Intestine 大腸
NanLü 南呂	10	*You* 酉	Kidney 腎	Lung 肺
WuYi 無射	11	*Xu* 戌	Pericardium 心包	Stomach 胃
YingZhong 應鐘	12	*Hai* 亥	Triple Heater 三焦	Urinary Bladder 膀胱

(The information in this chart is drawn from Wu, 2014)

Potential clinical uses

Five Phases

The above is a fairly cursory exploration of each of these numbers explored in either system. From the perspective of acupuncture, if the twelve channels can be overlaid on the twelve notes and then five-note scales generated from there, this could equate each of the five phase notes on each channel. Then each note could accord to a single point, giving very clear treatment strategies through simple musical compositions. What this could look like is creating the entire 12-note *huangzhong* scale and then building a *gong* scale with each one as the tonic.

It is important to note that the *huangzhong* note is fluid, and newly established with each emperor (Fisher 1973, 32). Since we do not have an emperor to accord with, we could consider the *huangzhong* to be C as it is the most common central note ("middle C," "Do" in the solfege system). This supposition is not certain and if one had an argument for a different *huangzhong*, one could certainly work from there. However, as the *huangzhong* is thought to be a universal force of sound, it would be most sensible to think of it as completely outside the realm of musical scales, and as it came through the emperor, it was able to take a sound form in the material world, as he was considered a god-manifest by the ancient Chinese. Because we have no divine figure that is thought to be channeling heavenly force into the world, it would be hard to think of us tuning to any single note. Furthermore, as East Asian medicine sees humans as microcosms of the universe, and medicine is working at the level of the individual, it makes more sense to tune to each individual, especially in the modern world where humans have a more individualistic outlook than in ancient China. Fundamentally speaking, the five phases themselves are primordial forces that are emanated down into the world (much like Socratic forms and planes). Likewise, nodal sounds (結聲 *jiesheng*) are primordial sounds emanating from the caverns of the five phases. However, in order to work with them in humans, they are translated into scale degrees. When we use a universal force in the human body it is no longer thought of in terms of its purest form, but is seen in context and conversation with other forces.

This would mean, working with the *huangzhong* as C, the *gong* scale on the gallbladder channel would accord with the most common version of that scale (the earth point as C, the metal point as D[15], etc). To work the liver channel the *gong* note would be C♯, the metal point would accord with D♯, etc. This would continue through the rest of the notes via the earthly branch/*DiaoSheng* association. Peter Moran has a detailed chart in his text giving the hertz values of all of these notes using the fifth and fourth ratios (3/2 and 9/8 from the tonic, respectively) to determine all of the values. He uses A440 as the *huangzhong* note. Note that he sees each of the five notes as a different pentatonic scale model rather than a scale degree, simply taking the same interval model, but starting from any one of the notes. This can be recreated with middle C as the tonic (261.626 hertz) to serve the model of this paper better, although again, the *huangzhong* of the individual would be a far better model than a more societal level *huangzhong* of C.

[15] It is important to note that this is untempered tuning based on the earlier explanation basing all notes on the perfect fourth/fifth interval. The base note is taken, raised by a fifth, then down a fourth, and so on, until these twelve are reached. While the Western notes may not be perfect equivalents to the Chinese *huangzhong* scale, they are approximate and are used to maintain simplicity for the reader.

Table 6. Showing hertz values of the *huangzhong* scale with C as the tonic, and then twelve *huangzhong* scales with each of the notes of the first scale serving as the tonic to determine the appropriate *huangzhong* scale for each organ/meridian system.

12 Lü	Ratio	From 261.626	From 279.383	From 294.329	From 315.877	From 331.120	From 353.425	From 372.510	From 392.439	From 419.074	From 441.494	From 471.438	From 496.681
Huang-Zhong 黃鐘	1/1	261.626	279.383	294.329	315.877	331.120	353.425	372.510	392.439	419.074	441.494	471.438	496.681
DaLü 大呂	2187/2048	279.383	298.345	314.305	337.316	353.593	377.412	397.793	419.074	447.517	471.459	503.435	530.391
TaiCu 太簇	9/8	294.329	314.306	331.120	355.362	372.51	397.603	419.074	441.494	471.458	496.681	530.368	558.766
JiaZhong 夾鐘	1968/1630	315.877	337.316	355.362	381.378	399.782	426.712	449.754	473.816	505.974	533.043	569.196	599.674
GuXian 姑洗	81/64	331.120	353.594	372.510	399.782	419.074	447.304	471.458	496.681	530.391	558.766	596.664	628.612
ZhongLü 仲呂	1771/1311	353.425	377.412	397.602	426.711	447.302	477.434	503.215	530.137	566.118	596.404	636.855	670.955
RuiBin 蕤賓	729/512	372.510	397.793	419.074	449.755	471.458	503.216	530.390	558.766	596.689	628.612	671.247	707.188
LinZhong 林鐘	3/2	392.439	419.074	441.494	473.816	496.68	530.138	558.765	588.659	628.611	662.241	707.157	745.022
YiZe 夷則	6561/4096	419.074	447.518	471.458	505.974	530.390	566.119	596.689	628.611	671.276	707.188	755.153	795.587
NanLü 南呂	27/16	441.494	471.459	496.680	533.042	558.765	596.405	628.611	662.241	707.187	745.021	795.552	838.149
WuYi 無射	5905/3277	471.438	503.435	530.367	569.196	596.663	636.855	671.246	707.157	755.152	795.551	849.509	894.996
YingZhong 應鐘	243/128	496.681	530.391	558.765	599.673	628.611	670.955	707.187	745.021	795.586	838.149	894.996	942.918

It is most likely that the most effective way to work with these would be to keep the *gong* note, and possibly also the *huangzhong* note, droning while playing over them, as music is contextual and notes without a relationship to one another are harder to give a meaning. This also makes sense in the sense of the diagrams of the five elements drawn with earth in the middle. Earth as *gong* is the point from which one goes in any direction. Thus, the rest of the pentatonics are meaningless without the tonic to give context. Although presuming the patient is tuned to the *huangzhong*, we could think that in a sense the interior is always droning that tone, setting that the entire set of *gong* tones for the twelve channels, so one could argue for playing them as a drone, or for the drone as unnecessary.

To work with these notes, it would make sense to correspond the playing of the note with how needles are moved in acupuncture. To tonify, slow, steady notes, at a comfortable pitch would make sense. For dispersion, faster, louder, more staccato notes would be appropriate. To circulate *qi* or blood, running through the scale repeatedly in the order the points appear on the channel would be most sensible. To calm, playing soft and slow, allowing notes to fade out would be best. These are things most musicians are likely to do intuitively, but these proposals more directly accord how one works with a channel with how one plays an instrument.

Below are the five phase points of each channel based on this idea, assuming the *huangzhong* is C. Obviously, if it was another note, it would need to be modulated. Note that this is an untempered scale.

As an example, to tonify the heart channel, as mentioned above, one tonifies Ht9, Lv1, and disperses Ht3, Kd10. A way this could be done would be to play F♯ with A♯ (that is the *gong* of the heart channel with the *jue* of the heart channel), then C♯ with F (the *gong* of the liver channel with

Table 7. Showing five phase points of each channel, with their associated note, based on a *huangzhong* of C

Channel	宮 *Gong* (Earth)		商 *Shang* (Metal)		角 *Jue* (Wood)		徵 *Zhi* (Fire)		羽 *Yu* (Water)	
Gallbladder	C 261.626	Gb34	D 294.329	Gb44	E 331.120	Gb41	G 392.439	Gb38	A 441.494	Gb43
Liver	C♯ 279.383	Lv3	D♯ 314.306	Lv4	F 353.594	Lv1	G♯ 419.074	Lv2	A♯ 471.459	Lv8
Lung	D 294.329	Lu9	E 331.120	Lu8	F♯ 372.510	Lu11	A 441.494	Lu10	B 496.680	Lu5
Large Intestine	D♯ 315.877	LI11	F 355.362	LI1	G 399.782	LI3	A♯ 473.816	LI5	C 533.042	LI2
Stomach	E 331.120	St36	F♯ 372.51	St45	G♯ 419.074	St43	B 496.68	St41	C♯ 558.765	St44
Spleen	F 353.425	Sp3	G 397.603	Sp5	A 447.304	Sp1	C 530.138	Sp2	D 596.405	Sp9
Heart	F♯ 372.510	Ht7	G♯ 419.074	Ht4	A♯ 471.458	Ht9	C♯ 558.765	Ht8	D♯ 628.611	Ht3
Small Intestine	G 392.439	SI8	A 441.494	SI1	B 496.681	SI3	D 588.659	SI5	E 662.241	SI2
Urinary Bladder	G♯ 419.074	UB40	A♯ 471.458	UB67	C 530.391	UB65	D♯ 628.611	UB60	F 707.187	UB66
Kidney	A 441.494	Kd3	B 496.681	Kd7	C♯ 558.766	Kd1	E 662.241	Kd2	F♯ 745.021	Kd10
Pericardium	A♯ 471.438	Pc7	C 530.368	Pc5	D 596.664	Pc9	F 707.157	Pc8	G 795.552	Pc3
Triple Heater	B 496.681	TH10	C♯ 558.766	TH1	D♯ 628.612	TH3	F♯ 745.022	TH6	G♯ 838.149	TH2

the *jue* of the liver channel), with a slow, building tone. Then follow with F♯ with D♯ (*gong* of the heart channel with the *yu* of the heart), then A with F♯ (*gong* of the kidney channel with the *yu* of the kidney channel) more staccato and forcefully. This could potentially be done with a C droning in the background, setting the *huangzhong*, though this would not be considered absolutely necessary.

If we want to accord with these two their Western equivalents, this would make the first into a strange set of intervals, the next into a C♯ major 7th chord (with the F as an E♯), the third into a C diminished (with D♯ and F♯ as E♭ and G♭), and the fourth into an F♯ diminished. If we combine the tonification notes, it would produce an F♯ major 7th (calling F an E♯), and combining the dispersing notes produces a D♯ diminished.

C and F♯ are a tritone apart from each other, making them contrary scales. This may make the C drone on the F♯ based chords slightly strange to the western ear. In terms of channels the tritone would be the same as the channel on the opposite side of the Yuan cycle[16].

This chart (Table 8) is the same as the Korean Sa'am five element acupuncture system chart that was presented earlier, with the notes of the appropriate *gong* scales substituted in for points. The note in parentheses is the gong note of the scale of the note being utilized. The numbers are the hertz values based on the *huangzhong* as 261.626 (middle C).

Seven Emotions

With the seven emotions system, one can create a treatment based on dispersing the organ that is associated with the excessive emotions and tonify that which is damaged by that excess. As the emotions relate to organs rather than channels, these need to follow the organ correspondences of the *DiaoSheng* system, rather than the channel correspondences, as listed in Table 5.

There are a number of ways one could think about approaching the seven emotions with ancient Chinese music

[16] The Yuan cycle is the Chinese clock, correlated with the earthly branches, that shows when *qi* is most prominent in any given channel. Simultaneous aberrations on opposite sides of the *yuan* cycle can be thought to indicate an issue with *yuan qi*.

Table 8. Showing Korean Sa'am five element acupuncture system chart with the correlated notes based on a *huangzhong* of C

Channel	To Tonify				To Disperse			
	Tonify		Disperse		Tonify		Disperse	
Lung	D (D) 294.329 (294.329)	F (F) 353.425 (352.425)	A (D) 441.494 (294.329)	C♯ (F♯) 558.765 (372.510)	A (D) 441.494 (294.329)	C♯ (F♯) 558.765 (372.510)	B (D) 496.680 (294.329)	F♯ (A) 745.021 (441.494)
Large Intestine	D♯ (D♯) 315.877 (315.877)	E (E) 331.120 (331.120)	A♯ (D♯) 473.816 (315.877)	D (G) 588.659 (392.439)	A♯ (D♯) 473.816 (315.877)	D (G) 588.659 (392.439)	C (D♯) 533.042 (315.877)	F (G♯) 707.187 (419.074)
Stomach	B (E) 496.68 (331.120)	D (G) 588.659 (392.439)	G♯ (E) 419.074 (331.120)	E (C) 331.120 (261.626)	G♯ (E) 419.074 (331.120)	E (C) 331.120 (261.626)	F♯ (E) 372.51 (331.120)	F (D♯) 355.362 (315.877)
Spleen	C (F) 530.138 (353.425)	C♯ (F♯) 558.765 (372.510)	A (F) 447.304 (353.425)	F (C♯) 353.594 (279.383)	A (F) 447.304 (353.425)	F (C♯) 353.594 (279.383)	G (F) 397.603 (353.425)	E (D) 331.120 (294.329)
Heart	A♯ (F♯) 471.458 (372.510)	F (C♯) 353.594 (279.383)	D♯ (F♯) 628.611 (372.510)	F♯ (A) 745.021 (441.494)	D♯ (F♯) 628.611 (372.510)	F♯ (A) 745.021 (441.494)	F♯ (F♯) 372.510 (372.510)	F (F) 353.425 (353.425)
Small Intestine	B (G) 496.681 (392.439)	E (C) 331.120 (261.626)	E (G) 662.241 (392.429)	F (G♯) 707.187 (419.074)	E (G) 662.241 (392.429)	F (G♯) 707.187 (419.074)	G (G) 392.439 (392.439)	E (E) 331.120 (331.120)
Urinary Bladder	A♯ (G♯) 471.458 (419.074)	E (D) 331.120 (294.329)	G♯ (G♯) 419.074 (419.074)	E (E) 331.120 (331.120)	G♯ (G♯) 419.074 (419.074)	E (E) 331.120 (331.120)	C (G♯) 530.391 (419.074)	E (C) 331.120 (261.626)
Kidney	B (A) 496.681 (441.494)	F (D♯) 355.362 (315.877)	A (A) 441.494 (441.494)	F (F) 353.425 (353.425)	A (A) 441.494 (441.494)	F (F) 353.425 (353.425)	C♯ (A) 558.766 (441.494)	F (C♯) 353.594 (279.383)
Pericardium	D (A♯) 596.664 (471.438)	F (C♯) 353.594 (279.383)	G (A♯) 795.552 (471.438)	F♯ (A) 745.021 (441.494)	G (A♯) 795.552 (471.438)	F♯ (A) 745.021 (441.494)	A♯ (A♯) 471.438 (471.438)	F (F) 353.425 (353.425)
Triple Heater	D♯ (B) 628.612 (496.681)	E (C) 331.120 (261.626)	G♯ (B) 838.149 (496.681)	F (G♯) 707.187 (419.074)	G♯ (B) 838.149 (496.681)	F (G♯) 707.187 (419.074)	B (B) 496.681 (496.681)	E (E) 331.120 (331.120)
Gallbladder	A (C) 441.494 (261.626)	F (G♯) 707.187 (419.074)	D (C) 294.329 (261.626)	F (D♯) 355.362 (315.877)	D (C) 294.329 (261.626)	F (D♯) 355.362 (315.877)	G (C) 392.439 (261.626)	D (G) 588.659 (392.439)
Liver	A♯ (C♯) 471.459 (279.383)	F♯ (A) 745.021 (441.494)	D♯ (C♯) 314.306 (279.383)	E (D) 331.120 (294.329)	D♯ (C♯) 314.306 (279.383)	E (D) 331.120 (294.329)	G♯ (C♯) 419.074 (279.383)	C♯ (F♯) 558.765 (372.510)

theory. One simple way is to see which organ is imbalanced and build that as the gong note of the scale, and another would be to take the imbalanced organ and put it in the corresponding note of the scale (so if the liver is imbalanced, take the *jiazhong* note of the *huangzhong* scale and build a heptatonic scale around that note in the *jue* position). The only way to be sure of efficacy would be through clinical trials.

Using the former C♯[17] and G serve as the *gong* of spleen, D♯ the *gong* of liver, F♯ the *gong* of heart, A the *gong* of lung, C the *gong* of kidney. Assuming the latter model was used, then C♯ would be the *gong* note for the spleen. D♯ would be the *jue* note of the liver scale, making B the *gong* note of that scale. F♯ as the *zhi* note of heart would put C♯ as the *gong* note. A as the *shang* note of lung would make G the *gong* note in that scale. C as the *yu* note of kidney would make D♯ the *gong* of kidney. There would also be the alternate spleen scale, in which G is the *gong* note (same as the lung). This alternate scale would be useful when thinking about the two emotions associated with spleen, worry and over-thinking. Worry having more relationship to the lung, and over-thinking having more relationship to the heart, would make it obvious which scale to use in either case. In addition, being that in order of note generation (via the ascending fifth, descending fourth), the latter spleen note would appear first (as it is the first fifth), it makes sense to make that primary and the other secondary in either model.

All that is missing is how to account for the scale for fright (驚 *jing*). To figure this out, we must decide upon an organ correlation for that emotion. The *LingShu* seems to make it most closely related to heart, as it leaves the heart without a place to lean and gives the spirit (神 *shen*) no place to land. There is also the possibility that the second note correlated to the stomach could account for it, by assuming the second notes for the two that have two notes (only stomach and spleen) can be thought to stand for these notes that fall outside the normal system.

[17] This is again assuming a *huangzhong* of C. If another *huangzhong* is used, these notes must shift accordingly.

The first model appears as follows:

Table 9. Showing the first model of working with the seven emotions and the Chinese seven-note scale.

	Gong 宫	*Shang* 商	*Jue* 角	*BianZhi* 變徵	*Zhi* 徵	*Yu* 羽	*BianGong* 變宮
Spleen 脾 1	G 392.439	A 441.494	B 496.681	C♯ 558.766	D 588.659	E 662.241	F♯ 745.021
Lung 肺	A 441.494	B 496.681	C♯ 558.766	D♯ 628.612	E 662.241	F♯ 745.021	G♯ 838.149
Kidney 腎	C 261.626	D 294.329	E 331.120	F♯ 373.510	G 392.439	A 441.494	B 496.681
Liver 肝	D♯ (E♭) 315.877	F 355.362	G 399.782	A 449.755	A♯ (B♭) 473.816	C 533.042	D 599.673
Heart 心	F♯ 372.510	G♯ 419.074	A♯ 471.458	C (B♯) 530.390	C♯ 558.765	D♯ 628.611	F (E♯) 707.187
Spleen 脾 2	C♯ 279.383	D♯ 314.306	F 353.594	G 397.793	G♯ (A♭) 419.074	A♯ (B♭) 471.459	C 530.391
Heart 心 or Stomach 胃	F♯ 372.510	G♯ 419.074	A♯ 471.458	C (B♯) 530.390	C♯ 558.765	D♯ 628.611	F (E♯) 707.187
	A♯ (B♭) 471.438	C 530.368	D 596.664	E 671.247	F 707.157	G 795.552	A 894.996

Second model (core note is bold):

Table 10. Showing alternate model of working with the seven emotions and the Chinese seven-note scale.

	Gong 宫	*Shang* 商	*Jue* 角	*BianZhi* 變徵	*Zhi* 徵	*Yu* 羽	*BianGong* 變宮
Spleen 脾 1	**G** 392.439	A 441.494	B 496.681	C♯ 558.766	D 588.659	E 662.241	F♯ 745.021
Lung 肺	G 795.551	**A** 441.494	B 496.681	C♯ 558.766	D 596.404	E 662.241	F♯ 745.021
Kidney 腎	D♯ 315.877	F 353.425	G 392.439	A 441.494	A♯ (B♭) 471.438	**C** 261.626	D 294.329
Liver 肝	B 505.974	C♯ 569.196	**D♯** 315.877	F (E♯) 355.362	F♯ 381.378	G♯ 426.711	A♯ 473.816
Heart 心	B 503.216	C♯ 558.765	D♯ 628.611	F (E♯) 707.187	**F♯** 372.510	G♯ 419.074	A♯ 471.458
Spleen 脾 2	D 298.345	E 337.316	F♯ 377.412	G♯ 419.074	A 447.518	B 503.435	**C♯** 279.383
Heart 心 or Stomach 胃	C 530.390	D 596.689	E 671.246	**F♯** 372.510	G 397.793	A 449.754	B 503.215
	E 671.247	F♯ 755.153	G♯ 849.509	**A♯** 471.438	B 503.435	C♯ 569.196	D♯ 636.855

In either of these two models, the use of the heptatonic gong scale would be to tonify that which is depleted and to disperse that which is excess. In a case of mania injuring the corporeal soul (魄 *po*), by the first model, with an F♯ droning, one would disperse C♯ and tonify G♯. By the second model this would be a B drone, dispersing F♯ and tonifying C♯. For fright, there would be four possible scenarios, dispersing the correlative *bianzhi* and tonifying that which is negatively affected, the heart. It could be an F♯ drone while dispersing B♯ and tonifying C♯, A♯ drone while dispersing E and tonifying F, C drone while dispersing F♯ and tonifying G, or E♭ drone while dispersing A♯ and tonifying B. All of this could be played over a general C drone as the *huangzhong*.

Conclusion

While it is hard to say for certain how the ancient Chinese music system worked, as there was no hertz measure, it is illogical to conclude that its pentatonic, heptatonic, or chromatic scales would be based on static Western notes that we are acclimated to. It is much more logical to view them as mutable scale degrees, which move in smaller (*Gongyin*) and larger (*DiaoSheng*) cycles. Via the Chinese numerological correspondences, we are able to determine their correspondences and determine the most logical way to apply them therapeutically. While this is certainly more complex than applying a single note to a single element, by musical principles notes by themselves do not have much meaning. Their significance comes primarily in context together with other notes. At this point this is all theoretical, but hopefully with clinical trial it could be determined whether this is a more viable and efficacious system of treatment.

Acknowledgements

Thank you to Gabriel Marin for being a constant resource of musical knowledge. Thank you to Adam Baritot for reading earlier drafts and the invaluable feedback. Thank you to Dr. Tim Klemt for being a sounding board for these ideas. And thank you to Dr. Stephen Jackowicz for infinite resource suggestions.

References

Brindley, Erica Fox. *Music, Cosmology, and the Politics of Harmony in Early China.* State University Press of New York, 2012.

陈应时 [Chen, Ying-shi]. "一种体系 两个系统:一论中国传统音乐理论中的' 宫调" ["On the GongDiao of Chinese Traditional Music Theory"], *中国音乐学 [Musicology in China].* 2002 (4): 109–116.

Fisher, Fred. "The Yellow Bell of China and the Endless Search," *Music Educator's Journal*, Vol. 59, No. 8. (April 1973): 30-33, 95-98.

Jackowicz, Stephen. "The Sound and the Fury: An Historical Analysis of Sound and Music in Traditional Chinese Medical Culture." Interdisciplinary Society for Quantitative Research in Music and Medicine 2013 *Proceedings Journal*. June 2014.

Kaptchuck, Ted. *The Web that has No Weaver*. McGraw-Hill Education, 2000.

Khan, Hazrat Inayat. *The Mysticism of Sound and Music*. Shambhala, 1996.

Larre, Claude, and Elisabeth Rochat. *The Seven Emotions*. Monkey press, 1996.

LingShu (靈樞). China, ~1st century BCE.

Middendorf, Ulrike. "The Sage without Emotion: Music, Mind and Politics in Xi Kang," *Philosophy and Religion in Early Medieval China*. State University of New York Press, 2010: 135-171

Modirzadeh, Hafez. "Aural Archetypes and Cyclic Perspectives in the Work of John Coltrane and Ancient Chinese Music Theory," *Black Music Research Journal*, Vol. 21, No. 1 (Spring, 2001), pp. 75-106.

SuWen (素聞). China, ~1st century BCE.

Willmont, Dennis. *Yin-Yang and the Inner Phases: Understanding the Five Phases of Acupuncture in Theory and Clinical Practice.* Willmountain Press, 2010.

Wu, Zhongxian, and Karin Taylor Wu. *TianGan DiZhi*. Singing Dragon, 2014.

Correlation Between Traditional East Asian Acupuncture Hologram Theory and Replicative Self-symmetry in Musical Scales

Raymond Hong

Abstract

Recurrent self-symmetry is observable throughout nature, seen in both the human body and many musical scales. The former has been used in what are called "microsystems" - mapping the whole human body unto a smaller part, such as hands, feet, face, etc., independently by many cultures around the world for thousands of years. This self-replication of the whole unto the part is also known as "hologram theory." Microsystems and hologram theory are the centerpiece to many systems of body-based traditional East Asian medicine, including massage, qigong (esoteric breathing exercises), tool-based treatment such as acupuncture, and body-based diagnostic measures (tongue/pulse/face/abdomen inspection).

Recurrent self-symmetry also manifests in music across many musical traditions. One of the most evident examples of self-symmetry in music is in the scales themselves, specifically the ratios between the frequencies of notes and intervals. Some ancient Greek traditions believed that the subjective idea of beauty arose from mathematical self-symmetry and replicative self-proportionality. It has been common belief and practice in many societies and eras that music can have beneficial properties for human health and human healing. This paper explores the concept of how the stimulation of one part of an organic system can have impact on the whole to restore balance or equilibrium. Further, the paper examines the correlation of unity between scalar parts of the system and its whole that equate to health and well-being.

Foreword

This is an exploratory paper, not based on any conclusive experimental research. It is an exposition of theoretical parallels and correlations, and not a definitive proof of a common link between musical scales and the role of self-similarity in human health.

The paper proceeds in three sections. First, an examination of these aforementioned concepts in music. Second, an examination of them in East Asian Medicine. Third and lastly, a correlative section that interrelates them.

A: Music

a1. Western scales - Pythagorean and Even-tempered

The Pythagorean scale ascribes ratios of whole numbers to the frequencies/wavelengths of musical intervals.[1] Herein where the mathematical relationship between the frequency and wavelength of a vibration is an inverse one, and one necessarily defines the other, I will refer only to frequencies for the rest of this paper.

The chords commonly used in Western musical composition, such as the major and minor triads, achieve harmonic ratios of their wavelengths. The ancient Pythagoreans believed, along with some modern scholars, that these integer ratios of the wavelengths are what lend music its symmetry, and from its symmetry, its beauty. There are also other musical traditions around the world, such as from ancient China and India, that sought to link the mathematical laws of harmonics and rhythms to the laws of nature and human well-being.[2] However, this paper focuses on the modern even-tempered Western scale and its Pythagorean ancestor.

As an interesting note, the outer human ear is noted by some to reflect a spiral derived from the mathematical "Golden Ratio" (based on the numerical constant now known as *phi*, or approximately 1.618), and the cochlea of the inner ear where sound waves are actually converted into neural signals is also shaped like a spiral, albeit one that does not perfectly conform to the golden ratio itself.[3]

The golden ratio and *phi* see a connection to the whole integer series, popularly known as the Fibonacci sequence

(1, 1, 2, 3, 5, 8, 13, 21, etc.), where the ratio between each two consecutive numbers approaches *phi*, as the sequence approaches infinity.[4] Most of the intervals in the Pythagorean tuning system can therefore be expressed in ratios of Fibonacci numbers.[5] Some of the intervals that do not initially appear to follow this pattern can be adjusted by *inversion* or going up or down an octave. Thereby the minor third's ratio of 6/5, if the higher pitch is lowered an octave, is divided by two and becomes 3/5; the major third's ratio of 5/4, if the higher pitch is raised an octave, is multiplied by two and becomes 5/2; the perfect fourth's ratio of 4/3, if the higher pitch is lowered an octave, is divided by two and becomes 2/3.

The original Pythagorean intervals lend themselves towards harmonics. Where the frequencies, or wavelengths, of two sound vibrations are ratios of whole integers to one another, their vibrations will coincide in regular periods.[6] In a Pythagorean major fifth, of a ratio of 3/2, between A4 and E5, with A4 set to 440 Hz, E5 would be 440 Hz × 3/2 = 660 Hz. That means that every second, the object resonating A4 would vibrate 440 times, and the object resonating E5 would vibrate 660 times. For every 2 vibrations, or waves, of A4, there would be precisely 3 vibrations or waves of E5. This produces a consonant effect in music. Many of the more consonant intervals and chords (simultaneously played notes) are in small integral ratios to one another, whereas the more dissonant intervals, such as the tritone (augmented fourth, diminished fifth) were, at best, ratios of large integers such as the Pythagorean 729:512 or the "just" intonation 25/18 or 45/32. Another oft-mentioned coincidence of numbering regarding the Pythagorean scale is that including both root and octave notes, the octave scale numbers eight notes, the interspersed semitones number five, the full number of tones number thirteen - five, eight and thirteen being Fibonacci numbers.

Let us examine the Pythagorean ratio of the frequencies of the major and minor triads, which are two of the most commonly-used chords in Western music. These are, respectively, 4:5:6 and 10:12:15.[7] Using inversion, one can divide 4 and 6 each by 2 if one lowers those notes by an octave, and result in 2:5:3 - a reoccurrence of Fibonacci. If one looks at the wavelengths of the minor triad instead of the frequency, one will use the reciprocals or inverse of the ratio 10:12:15, or 1/10 : 1/12 : 1/15. Multiplying each one by their lowest common multiple, 60, one obtains the ratio of 6:5:4 for wavelength - which by the same process of octave inversion above can yield 3:5:2 - also Fibonacci numbers.

There is another interesting parallel between *phi* and the Fibonacci sequence and musical interval theory. *Phi* is an irrational number not representable as a ratio of any integers, whereas the Fibonacci sequence is an infinite series of integers, the ratio between any two consecutive elements of which approaches *phi* (and the golden ratio) as the sequence approaches infinity. The Pythagorean tuning system, and many other antique tuning systems around the world also based on ratios of whole integers, such as that in China and India, suffers discrepancies, especially as music intervals diverge from a "root" octave, expressed in the Pythagorean comma: "the difference between twelve justly tuned perfect fifths and seven octaves... expressed by the frequency ratio 531441:524288 (23.5 cents)".[8] This discrepancy was resolved in the 16th century with the logarithmic scale of equal temperament which became the most commonly-used scale in modern Western music.[9] Equal temperament, by the nature of exponents and logarithms, is irrational and cannot be expressed as ratios of whole integers. The Pythagorean ratios can be seen as approximating or approaching the intervals of this equal-tempered scale. Conversely, the intervals of the equal-tempered scale, though they are an irrational logarithmic scale, approach the simple ratios of the Pythagorean scale that lend themselves to harmonic frequencies. Is it the fact that the equal-tempered intervals approach rational harmonic frequencies, or the fact that the antique Pythagorean tunings approach that of the ultimately equal logarithmic temperament, that is what gives music its subjective beauty in the human experience?

a2. Rhythm

Two elements common to almost all music include a regular scale and a regular rhythm. The meter of such rhythm can vary in number, the most common in Western music over the years being 4 beats per "measure" or phrase, or 3, or 8 and 6, which are in effect, multiples of 4 and 3 by 2. The rate or pace of such rhythm can also vary and is an integral aspect to music around the world.

a3. Consonance and Dissonance

The duality of consonance and dissonance, and the interplay thereof to create certain observed emotional affects in music, is a well-known topic in music theory. The following quotations describe this duality. As stated by Tom Sutcliffe:

> It works on the assumption that consonant sounds are combinations of notes (intervals) that sound pleasant or melodious together whereas dissonant intervals sound harsh or unpleasant together.
>
> Another way to express this is to say that consonant intervals create a feeling of stability whereas dissonant intervals create a feeling of instability that needs to be resolved. It is important to say that ideas of what constitutes consonance and dissonance have varied to some extent over time and vary from individual to individual. Also, it is important to say that most people perceive not just a simple binary division but a spectrum of consonance and dissonance varying from the most consonant intervals to the most dissonant intervals. [10]

Or as described by Roger Kamien, in "Music: An Appreciation":

> An unstable tone combination is a dissonance; its tension demands an onward motion to a stable chord. Thus dissonant chords are "active"; traditionally they have been considered harsh and have expressed pain, grief, and conflict. [11]

These are the subjectively-derived adjectives for the effect, or rather emotional *affect*, of the pain, conflict, or otherwise tension built by dissonant chords and the natural yearning of the human listener for resolution towards a consonant chord.

An example is Ludwig van Beethoven's Piano Sonata No. 14 in C♯ minor "Quasi una fantasia", Op. 27, No. 2, popularly known as the Moonlight Sonata, Movement I, "Adagio sostenuto." The emotional climax and the most core-wrenching moments of the piece occur at the C♯ - D♮ - C♮ - C♯ tones in measures 52-54. As described by Pierre Beaudry:

> The entire movement is uniquely pivoting around the two painful intervals between Bb-B-C in the first case (measures 16-18), and C-C#-D in the second case (measures 52-54).[12]

There is a particularly sharp and intense pain elicited in these measures, albeit gradually built up to by the preceding body of the piece. It is then gradually or "gently" resolved and "mollified" in the following measures 55-60.[13]

B. Medicine

b1. "Microsystems" or hologram theory in traditional East Asian Medicine

The concept of a "map" is a scalar or non-scalar depiction of a larger object onto a smaller one. The theory of mapping the physical structure of the human body, known as "holographic theory" or "microsystems" has been present in various medical systems around the world,[14] such as in ancient China, Egypt, and North America.[15] This is but one concept of many in physical therapeutic practices around the world. For the purposes of this paper, the focus is on the physical therapeutic systems of the East Asian sphere, also known as the "Sinitic Sphere" - the region of East Asian that has historically been heavily influenced by Chinese traditional culture, generally corresponding to modern day China, Korea, Japan, and Vietnam.

One of the best-known traditional tool-based therapies from this region of this world is acupuncture, while there are other branches of traditional East Asian Medicine. One of the best-known tools used in acupuncture is the *hao zhen* or filiform needle, although this is just one out of many dozens of various tools used together in the same discipline. One of the constants underneath such variety in acupuncture modality and technique are the physical maps used with such modalities and techniques. Not every acupuncture theory or protocol can have a clean correspondence with microsystems, but microsystems are a frequently recurring idea within the broader realm of acupuncture from ancient times to recent developments in the modern day.

As a brief overview, the field of traditional East Asian Medicine is broadly categorized into four different independent and sometimes interrelated disciplines: acupuncture, moxibustion and tool medicine; herbology and diet; physical manipulation and massage; and *qigong* or esoteric breathing and moving exercises.

Many of the uses of the regular 359

acupuncture points may be theoretically connected back to microsystem theory. For instance, one indication for the point Du 26 *Shui Gou* on the frenulum of the upper lip is indicated for acute back pain, and that area happens to correspond to the lower back/abdomen of the torso section in one facial microsystem of the human body. Kidney 6 *Zhao Hai*, inferior to the medial malleolus of the tibia or just below the inside ankle, is a point used to strengthen Kidney Yin, believed to be closely linked to the reproductive system, can be seen as "inside" the lower pelvic region of the body where the foot proper is a microsystem for the body. This same area is the correspondence area for the uterus/prostate in foot reflexology.

There are a large number of coexisting and even competing traditions of microsystems. Pulse diagnosis, of which there were once many sampling spots on the human body but now only the wrist radial pulse is generally practiced, can be traced at least to the Han Dynasty around 2,000 years ago. Within the radial pulse itself, there is a homunculus - a representation of a small human being.[16] The homunculus is also represented in the various schools of abdominal diagnosis, in tongue diagnosis, and in the face, scalp, occiput, throat, forearms, lower legs, paraspinal muscles, and sacrum. Auricular or ear acupuncture is a relatively recent 20th century development based upon these same principles, as are two different and competing schools of Korean Hand Acupuncture (*Soojichim*) and Hand and Foot Acupuncture (*Sujok*). Among non-East Asian microsystems, the entirely foot and ankle-based school of reflexology,[15] and the eye-based diagnostic system of iridology discovered in 19th century Hungary,[17] bear mentioning. Many of these microsystems map onto physically discrete structures, often delineated by individual bones or bone groupings.

These microsystems are used both diagnostically and therapeutically, and potentially in the same stroke. Token palpation or observation of areas of any given microsystem can serve in diagnosis of the human body *in totum*. Diagnostic relevance attaches to areas that elicit pain in reaction to palpation, or demonstrate excess *rubor* or inflammatory discoloration in reaction to palpation, or both. The touch or pressure used in palpation is itself a massage technique, and so is already one therapeutic venue, as are the other tool-based therapies including needling. The therapeutic response is often immediate, as the whole now starts to reflect the change made to the part which once reflected the injurious or pathological change made to the whole.

It is a traditional concept in medicine, not just Chinese but also Greek, that the feeling (*pathos*) of the body should be as one unified and interconnected whole (*ortho* for "same"). In other words, the ancient ideal of health can be defined as being orthopathic, as opposed to being heteropathic, or feeling like part of the body is separate, or suffering separately, or having a separate experience as a result thereof.[16] Perhaps one mechanism of microsystem theory is that treating those microsystems reinforces the connection between the parts back to the whole, and to one another.

The Pythagoreans and Euclideans of ancient Greece saw the recurrence of the mathematical golden ratio not only in music, but also in the dimensions of the human body.[18] This ratio recurs in many locations throughout the body, including the limbs, hands, feet, and head, which are the very same locations and delineations of many of the microsystems mentioned above. That the mathematical, anatomical, and medical correspondences converge can be seen as coincidence, or beyond coincidence. In the case of the latter, where the golden ratio in the human body was seen in the West mostly as an aesthetic or merely curious phenomenon, the practice of microsystems brings the golden ratio a functional dimension as well.

Another geometric shape intimately related to the golden ratio and *phi* is the five-pointed unicursal star or pentagram. This shape is an expression of the concept of Five Phases, also rendered Five Elements, which is among those central to many traditional Chinese fields of study, including medicine. Together with a circle connecting the five corner vertices, forming a pentacle, each of those five corner vertices now share a connection or defined relationship with every other one.

At a glance, the five-pointed star is as an abstraction, albeit imperfect, of the human figure: one head, two arms and two legs, making for a central torso with five protrusions. Each of the hands and feet feature five digits, each becoming in turn another imperfect, scaled-down five-pointed star. The head itself presents a different challenge, with no obvious set of five anatomical protrusions. The ancient Chinese have a model of five senses that open up to the face: sight, taste, touch, smell, and hearing. The sense of touch is of course not restricted to the face alone. The "orifices" through which these senses are perceived are accordingly the eyes, tongue, mouth (lips and body flesh in general), nose, and ears.[19]

Consider another schema of mapping the human body - specifically, the axial portion, axial skeleton, or the torso and head. The human axial skeleton consists of a pelvic girdle, joined by the single lumbar spine to the thoracic cage, joined by the single cervical spine to the skull. This tripartite division is more obviously visible in insects, one of whose determining characteristics is the presence of a discrete head, thorax, and abdomen. This tripartite division of human and insect morphology are compared in Figure 1. The pelvic structure in humans is the thickest or widest and heaviest section, followed by the thorax, followed by the head or skull. The diagram is adjusted to better reflect a more structurally-intuitive parallel between the human skeleton and the insect exoskeleton by virtue of the width of the body parts and spacing of the axial joints. What is conventionally called the "abdomen" in the insect herein correlates with the pelvic girdle in humans. The relatively narrower abdomen and lumbar spine in humans is correlated with the narrow joint in insects between their own thorax and "abdomen."

Each of the four limbs each has three principal divisions. The arms divide

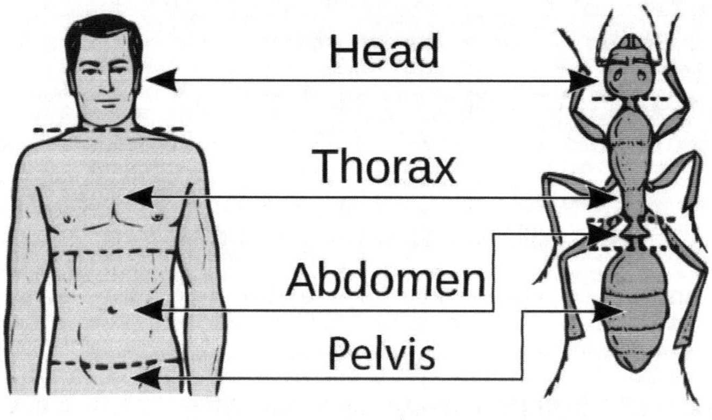

Figure 1. Diagram of the tripartite division of the human axial skeleton and insect morphology

into the upper arm, forearm, and hand/palm, while the legs divide into the upper leg (thigh), lower leg, and foot. The limbs are thickest and heaviest in their proximal sections (upper arm/leg), thinnest and lightest in their most distal sections (hands and feet), and somewhere in between in thickness and mass in their middle sections (forearms and lower legs). Determining a similar pattern for the skull itself again presents difficulties, which remain unresolved. Perhaps the skull can be divided based upon the orifice cavities: the mandible (lower jaw), separate from the maxilla and zygomatic bones (upper jaw and cheekbones), separate from the "neurocranium" (colloquial: "braincase") by the cavities comprising the eyes and nasal cavity, consisting principally of the frontal, parietal, temporal, and occipital bones.[20]

Following this schema, it follows that each of the four limbs (and the head, if the above postulation holds true) is a smaller reiteration of the torso. The upper arms would be analogous to the pelvic region, the elbow joints analogous to the lumbar region/lower back, the forearms analogous to the thoracic region, the wrists analogous to the cervical spine/neck, and the main body of the hands (just the "palm," excluding the digits/fingers) analogous to the head. The upper legs/thighs would be analogous to the pelvic region, the knee joints analogous to the lumbar region/lower back, the lower legs/shins/calves analogous to the thoracic region, the ankles analogous to the cervical spine/neck, and the main body of the feet, excluding the digits/toes, analogous to the head.

There are two notable items. First, one of the common axioms of traditional East Asian Medicine is that the kidneys govern the lower back and the knees, a relationship played out by this correlation.[16] Second, Western anatomy recognizes the posterior calf flexor muscles, the soleus and gastrocnemius, as "... responsible for pumping venous blood back into the heart from the periphery, and is often called the skeletal-muscle pump, peripheral heart or the sural (tricipital) pump." [21] The human heart develops from artery-like primitive endocardial tubes,[22] and the peripheral circulatory system of arteries and veins is structurally contiguous and similar to that of the heart ventricles and atria. The deep center of the calf, which in this instance refers to the deep soleus muscle, corresponds in this microsystem to the heart proper.

From this particular schema, one can determine the schematic origins or relationships of several commonly used acupuncture points or zones. Large Intestine 4 *He Gu* in the middle of the fleshy aspect of the dorsum of the hand between the first and second metacarpals is a "command point" for the face and mouth, while the opposable thumb can be seen as a kind of "chomping mouth" or "jaw" for a stylized head represented in the hands. Lung 9 *Tai Yuan* on the radial aspect of the wrist is a source or 'strengthening' point for the Lung meridian, used in some systems to address early colds and open restricted nasopharyngeal tracts, can be seen as corresponding to the base of the throat. Pericardium 6 *Nei Guan* in the palmar aspect of the forearm, approximately two inches proximal to the wrist in average adults, is a command point for the heart, chest, and epigastrium. (Urinary) Bladder 40 *Wei Zhong* in the middle of the popliteal fossa or crease on the back of the knee is another command point, popularly known for general back pain, but in some interpretations considered more appropriate for lower back and lumbar pain, while (Urinary) Bladder 58 *Fei Yang* near the center of the calf (back of the lower leg) is more often used for midback pain. Yet another Bladder channel point used for spinal or paraspinal pain is (Urinary) Bladder 60 *Kun Lun* posterior to the lateral malleolus on the ankle, for neck pain. Although standard acupuncture curriculums do not teach this usage, the author has personally used and 'rediscovered' this point usage for neck pain using the above schema as one contributing reason, together with five element theory. In this usage, the author’s clinical work supports greater efficacy in deviating from the standard textbook definition of the point location in the center of the fossa of the lateral malleolus, instead opting to needle more distally/inferiorly near the insertion of the Achilles tendon into the calcaneus, which is usually more tender. Lastly, the point St 38 *Tiao Kou*, otherwise "empirically" used for shoulder issues - shoulder pain, adhesive capsulitis or "frozen shoulder" and without explanation, is located on the "torso" of this stylized model of the human body.

Figure 2. Model of human body with scalar self-similar tripartite divisions

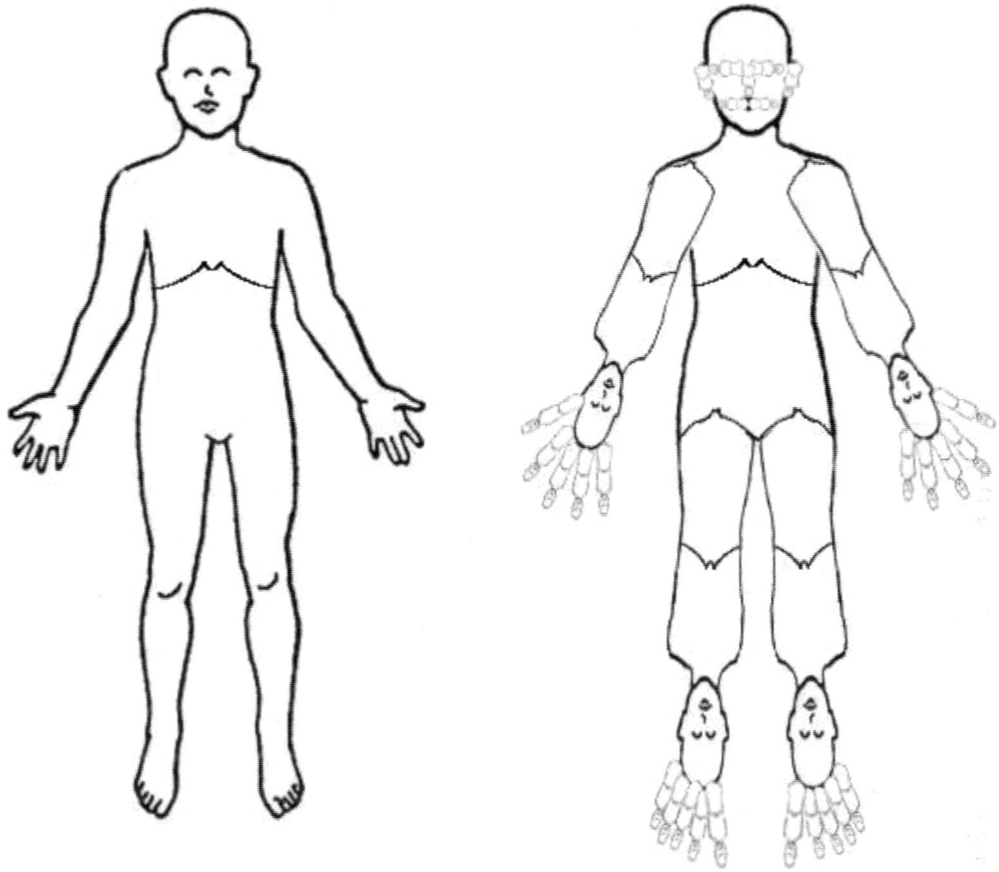

Continuing the pattern of scalar self-similarity, each of the fingers and toes can be modeled in this same manner as self-contained complete maps of this tripartite division of head, cervix/neck, thorax, lumbar, and pelvis. This model is illustrated in Figure 2. Four out of five of the digits on the hands and feet each have three phalanges or bones. The thumbs and large toes present another deviation and challenge, themselves having only two. This discrepancy in number of phalanges happens to be one of the reasons that Korean Hand and Foot Acupuncture (*Sujok*) founder Park Jae Woo disagreed with and ultimately split from his one-time teacher and founder of Korean Hand Acupuncture (*Soojichim*), Yoo Tae Woo.[16]

As for the head, there are microsystems for all of the orifices mentioned; the eyes form the basis of iridology/ocular acupuncture; the tongue is the basis of tongue inspection/diagnosis; the lips and the nose each have microsystems; and the ears have multiple auricular systems of acupuncture. Furthermore, there are other microsystems, including the face, forehead, eyebrows, parietal scalp, centerline of the scalp, occiput, forehead, and teeth. In a fashion similar to the fingers and toes of the hands and feet, there are smaller reiterated images of the human body on the distinguishable features of the face, numbering five.

Three distinct levels of scalar, albeit imperfect and non-proportional, self-similarity or reiteration in the human body have been listed: the main pelvis, thorax, and torso (joined by the lumbar and cervical joint regions), from which extend the four limbs each divided into three parts, and from each of which extend the twenty digits, most (sixteen) of which extend into three phalanges. One can therefore schematically represent the human body in a fractal based on the five-pointed star with the rule being replicating a smaller five pointed "child" star at each of the vertices of the larger "parent" star, and the examples listed here apply for three full iterations, illustrated in Figure 3.

The author has not yet encountered specific microsystem traditions based principally on the whole-limb schema but has learned of many traditions that employ different microsystem schema on different sections of the limbs. Some Taiwanese-based systems model the entire body on the forearm with hand and the lower leg with foot, such as the Master Tung's school, as well as on the upper arm and thigh alone, as is used in one of eight correspondence zones in Qingcheng Mountain Tuina 'massage' or body therapy, another traditional East Asian therapy outside that of acupuncture.[16] Some

Figure 3. One example of a fractal five-pointed star to three iterations

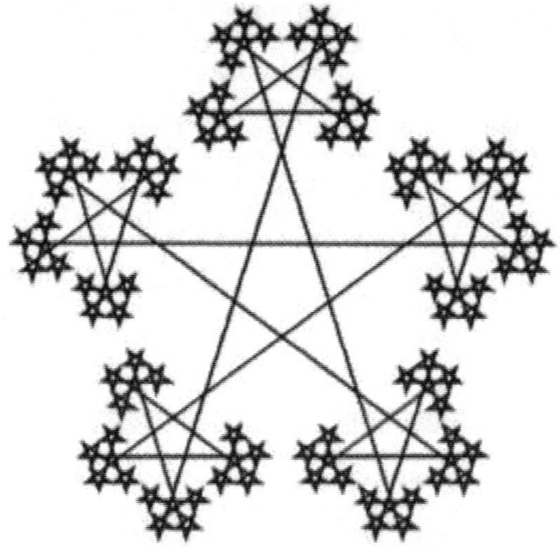

theories will even reverse the direction of the projection of the human body so the "head" is located on the proximal aspect of the map, rather than distal, and the "feet" vice versa.

b2. Binary system of treatment technique in Korean Sa'am Five Element acupuncture

Among the many styles of acupuncture practiced in the countries of the Sinitic sphere, one draws its roots from a 16th century Korean monk named Sa'am. His method is based on the five elements or phases mentioned before and has come to be known as Sa'am Five Element Acupuncture. One interesting set of rules in his method - not directly based on the five elements themselves - is the predication of "tonification" (or 'strengthening') and dispersion/sedating (or 'reducing') technique based upon five binary parameters: hand/foot (or up/down), left/right, yin/yang (or palmar/dorsal, or anterior/posterior), A.M./P.M. time of treatment, and male/female sex of the patient.[23] While there are a myriad of needle manipulation techniques in many other acupuncture traditions, there are only two in this particular one, tonification or dispersion.

Based upon the Sa'am determination of technique, "switching" any one parameter will necessitate the switching of the technique between tonification and dispersion. Switching a second parameter will accordingly switch the technique, incidentally, back to the original one. Switching a third parameter will switch the technique yet again, and so forth.

If one discounts the latter two of the five parameters, A.M./P.M. time of treatment and sex of the patient, then one is left with three geometrical binary parameters for determining needle technique on the human body. If one employs the traditional Chinese anatomical model for the human body, where the arms are abducted or raised, instead of adducted or lowered as in the Western model, then these three geometrical parameters more neatly correspond the three axes of dimensions, or three anatomical planes: transverse/horizontal, sagittal/longitudinal, and coronal/frontal/lateral. If one looks at the relationships of just two of these dimensions, up/down and left/right, one

Figure 4. Sa'am Five Element Acupuncture Binary Rules

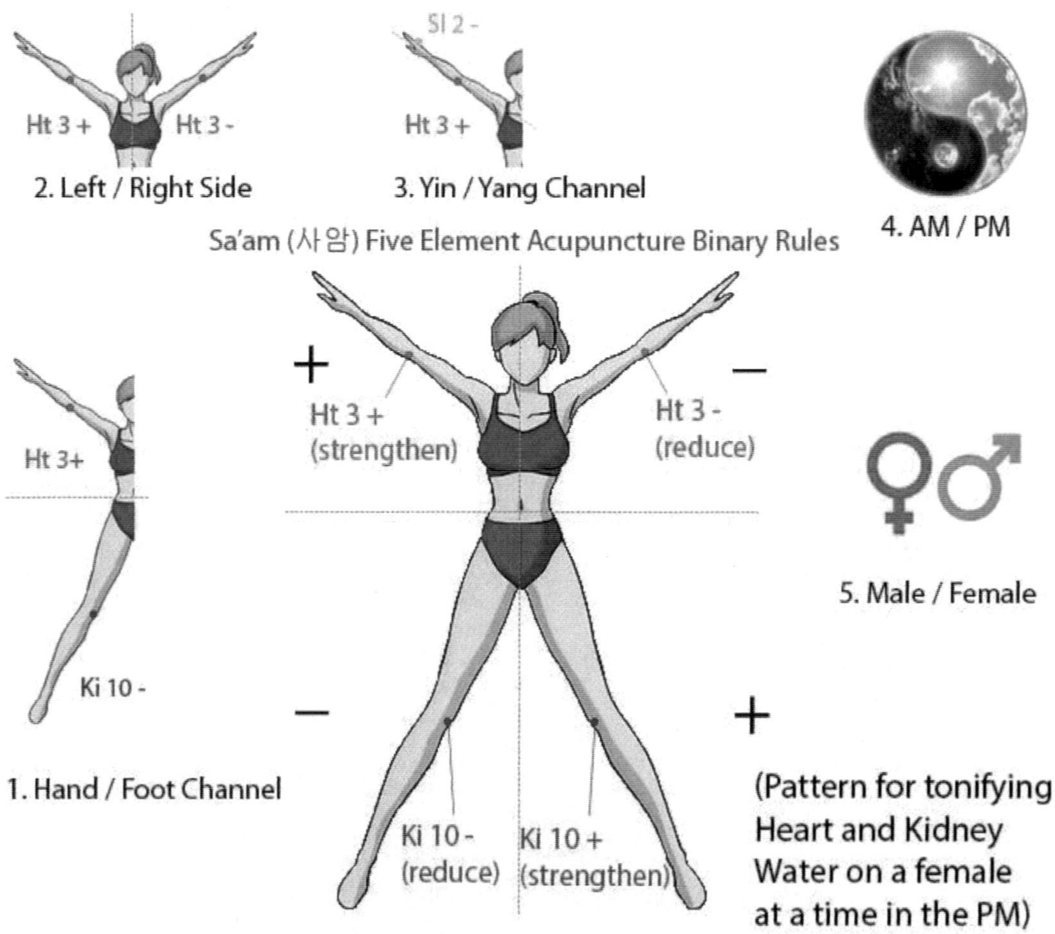

can divide the body into four quadrants. This relationship is illustrated in Figure 4, with the time of treatment in the P.M. and female sex of the patient. The other five binary parameters are displayed around the central figure, with parameters number 1, 2, and 3 illustrating the axes of reflection.

The result on the two-dimensional quadrant of the human body indicates a cross-wise relationship. If one were to tonify a point on a palmar (**anterior**) aspect or *yin* channel on the **upper right** side of the body (hand/arm), then instead of tonification, one would sedate the corresponding channel/point on the *yin* **lower right** side of the body (foot/leg). Similarly, one would also sedate the corresponding channel/point on the *yin* **upper left** side of the body (opposite hand/arm), in a kind of mirror reflection.

However, because two parameters are reversed when it comes to the *yin* **lower left** side of the body (opposite foot/leg), two mirror reflections causing the image to be "correct" again, one would tonify that corresponding channel or point. The relationship looks like an "X."

The cross-wise or "X" relationship occurs not only in this school, but in quite a number of other schools of traditional East Asian Medicine, martial arts and internal/external cultivation practices as well, including in the famous "Master Tung's" school of acupuncture and in the "Liu He" or "Six Harmonies" practice of *qigong*.

In embryonic development, the human (or any other animal) embryo begins as a single cell, which then undergoes mitation or binary cellular division. This division occurs along a single plane, and every structure divides and multiplies along that plane creating an identical copy, but also a mirror reflection of the original cell. This binary division occurs twice more before the cell first becomes a "blastocyst," beginning to adhere tightly (called "compaction") and this blastocyst has ideally eight cells - reflecting the divisions of the three dimensional or anatomical planes. If one follows the mirror reflection of cellular binary fission, these cells will exhibit the same patterns of symmetry exhibited in the binary rules determining technique in Sa'am. This is illustrated in Figure 5, where the crown denotes the first or original cell.

Figure 5. Theory of mirror reflection in binary cell division in embryonic development

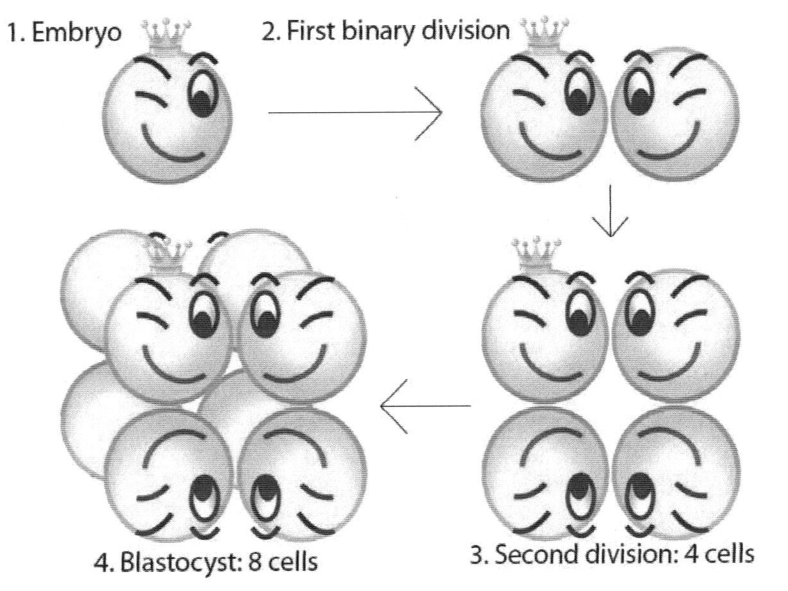

Where it is easy to see where one can obtain mirror, or bilateral/quadrilateral, or biplanar and ultimately triplanar symmetry in a relatively small blastocyst or embryo of just eight cells, achieving the same exacting level of symmetry across a human body numbering trillions of cells seems to require an altogether different mechanism of organization. As of yet, this enduring mystery of external bilateral symmetry in humans and many other animals, or any symmetry in lifeforms in general, remains unanswered. The closest any researcher has come to the molecular or chemical source of this symmetry is finding the "Pax" gene and protein in 1998, the expression of which suppresses the development of symmetry in the embryos of many orders of animals (including humans) to allow for the development of asymmetrical internal organ structures.[24] This discovery still does not explain exactly what higher order process is controlling the expression of this symmetry-suppressing Pax gene and protein, nor does it determine the fundamental process that allows for external bilateral symmetry across a wide expanse of the animal kingdom from small invertebrates like ants all the way to blue whales. Bilateral symmetry, and higher orders of self-symmetry, remain unexplained phenomena that are simply accepted in ancient traditions, including the ancient Chinese and the Pythagoreans.

The crosswise relationship mentioned above connects naturally with the concept of quadruped gait, which is practiced by most other closely-related species of the animal kingdom. Although human beings are bipeds, we are descended from quadrupeds, and human infants begin motion by crawling from about 6-9 months to about 12 months of age. The walking gait of most quadrupeds, such as elephants or horses, progresses in a cross-wise or "X" relationship - similar to that reflected in the cross-wise or "X" relationship, like the one described above, which is physically the most stable.[25] If one takes into account a third additional plane of movement, up and down (of lifting and lowering each leg), one now can make a connection between the three "yin and yang" or geometrically polar relationships employed in Sa'am Five Element Acupuncture and the triplanar or three-dimensional sequence of quadruped walking.

b3. Stretching, physical therapy and the human mind

The correlation between muscular or general physical tension and emotional tension is known[26] and employed in many fields of therapy, such as acupuncture, traditional East Asian Medicine in general, Western massage therapy, and especially the field of myofascial release (MFR).[27] From the website of the founder of MFR, John F. Barnes, fascia is defined as a "liquid crystalline matrix" that "in the normal healthy state... has the ability to stretch and move without restriction. When one experiences physical trauma, **emotional trauma**, scarring, or inflammation, however, the fascia loses its pliability. It becomes tight, restricted, and a source of tension to the rest of the body." [28] The emotional link to physical pain has been known in traditional East Asian Medicine for thousands of years, as mind and body are not considered entirely separate, but rather interconnected.

A close analog in traditional East Asian medicine to the emotional and physical qualities of this concept of fascia is *xue* 血, broadly translated as "Blood." Broadly translated, because the classical Chinese term carries multiple layers and facets of meaning beyond that of the inert modern Western concept of liquid blood outside of and divorced from the human body, and hence why "Blood" is capitalized in English-language acupuncture terminology to distinguish it from the Western concept of "blood." *Xue* Blood carries with it nuances of function and dynamism. *Xue* Blood in traditional East Asian thought was considered to be the "matrix of the mind" or the substrate of consciousness.[16] Fixed, sharp pain and areas of musculotendinous or metabolic restriction were considered *xue yu* or "Blood stagnation." Massage and tool medicine like acupuncture and moxibustion employ techniques to course or unblock such Blood stagnation. The tradition of Buddhism, a foreign import into China, found an affinity for tools that affect the Blood and its circulation, as Buddhism placed the *xin* 心 Heart (the "master" of the Blood in ancient Chinese thought, and also synonymous with the mind) at the center of their focus, and such tools like moxibustion, the burning of dried and shredded *artemisia argyii* or mugwort 'wool' historically spread with temple culture.[16]

Various traditions of physical stretching movements, including some traditional Chinese martial arts, other Chinese "self-cultivation" practices broadly grouped under the modern umbrella term "*qigong*," and the Indian tradition known as yoga, all integrate a body-mind connection and a venue of therapy for mental and emotional concerns. The ancients as well as modern schools of physical healing arts find that the very act of stretching and opening areas of myofascial restriction seems to have the effect of releasing emotional tension and blockage.

C. Music & Medicine

c1. The human waveform

It is a fundamental principle of traditional East Asian medicine to bring the body into symmetry and balance represented by the numerical concepts of yin and yang or the five phases. A symmetrical body is a healthy body, and an asymmetrical body is necessarily an unhealthy one. The very fact that an injury or disease in one part of the body reflects in symptomatic corresponding points on the various microsystems of the body is itself a display of the assertion of self-symmetry.

The usage of the microsystems is based on a general topographical congruence of each microsystem to the layout of the body proper. Granted, there are no parts of the body that precisely mirror the shape and proportions of the human body, and there is some "bending" or "stretching" and sometimes rotation of the layout, but the relative location of the main constituent areas, such as the head, lungs, or hip, are generally the same.

The presence of an injury or disease in one part of the body implies a breakage or a deviation from the normal symmetrical shape or function of the ideal state of the human body. Microsystem therapy restores the body's structural symmetry and functional relationships. In microsystem traditions that do not use separate tools but rather forms of massage, such as reflexology or Qingcheng Mountain Tui Na, it is various techniques of massage that are used to course or open the very blockages or restrictions in the tissue described above. In microsystem-based schools of acupuncture, tools including needles or stimulatory pellets and magnets are used to the same effect.

On a conceptual basis, perhaps what is occurring is not a direct connection or stimulation of the target body part through the corresponding microsystem point or area, but rather a restoration of the congruence and health of that microsystem. The restoration of the congruence of that microsystem in turn reestablishes its healthy relationship to the whole. In a sense, the healthy state can be seen as "consonant" while the diseased state can be seen as "dissonant." As described before, a healthy whole human living a healthy whole human existence can be described as "orthopathic." [16] The very sensation of a part of the body having a separate, or non-congruent, or dissonant experience from the rest, would be the opposite, or "heteropathic" or having the sensation of being separate and disconnected.

An important point to make here is that the body is not *truly* developing separate existences or entities. The aching finger or stomach will not spontaneously divorce itself and cleave off from the rest of the body. Rather, they are having a *sensation* of being separate or having a separate *experience*. This occurs in a grouping of musical notes that are dissonant, or a grouping of musicians or choir singers where one member plays or sings off key or off tempo. This creates a "clashing" resonance that does not concord with the rest, which would otherwise be perceived as a harmonious and unified singularity. The healthy human body and the healthy human experience should be exactly that - a singular existence. This concept of embracing and experiencing the singular is one of the primary goals of many traditional spiritual traditions, and in many of those traditions, having a singular experience ultimately with the greater world outside the limits of one's mind and body. As far as traditional East Asian medicine goes, suffice it to limit our focus to the patient suffering the lack of a singular experience within the limits of their own mind or body.

Some musical traditions, such as Confucian bell or Chinese pentatonic theory based on the five phases, or Hindustani and other South Asian raga traditions are highly systematized and focused towards healing specific health and emotional concerns. This paper does not examine these other specialized medicine-oriented musical traditions but draws a connection between the music based on the Pythagorean scale, which generally falls in what is recognized as the "Western world" or music in the Western cultural sphere.

Music in Western culture does not target specific diseases or body locations like microsystem therapy does. Music in general is a whole experience, that can be enjoyed (or not, as preferences are subjective) by either the healthy or the sick. The patient with the specific disease may still have their disease after listening to a song or having a musical experience. However, they may also experience a temporary improvement or cessation of their symptoms in the short term, and a generalized trend back towards wellness in the long term.

This experience of music encompasses the rational and harmonic congruence of the intervals and chords that underpin Western melody and harmony. This is an experience of resonance that follows rules of mathematics that also underpin the human body. This leaves a question, in the absence of a definitive answer: could the human body's fundamental operation be underpinned by a rational or fractal vibration?

If the body is based upon such a vibration, or dynamic system of multiple changing vibrations, then localized injury or disease can be modeled as a dissonant obstruction in the field of vibration, the practice of microsystem therapy on correspondence areas can be modeled as a disruption to the dissonant obstruction, and the process of healing a local injury or disease can be modeled as the reassertion of the body's proper/orthopathic/*zheng*-正-"upright" waveform back into that area from the periphery. The idea of the "obstruction" can be separate from the idea of the "dissonant vibration" caused by that obstruction, where if we were to use the analogy of a pond or ripple tank, the presence of a rock or other obstruction in the baseline wave has a distorting effect on it, such that it may generate a different wave. Such wave distortion may take the form of refraction, reflection, scattering, diffraction, absorption, or any combination of the above.[29] In this model, the generation of such a new, dissonant wave can very well attract the attention of one's mind and cause it to seem like a "separate" or "disconnected" existence.

It is important to note here that just like many practitioners of medicine or "healing arts" profess they are not 'healers' but rather facilitators of the body's innate ability to heal, the use of microsystem therapy does not directly "heal" or "correct" the corresponding area of the body. If the "vibration" model holds true, then the therapeutic technique applied to the microsystem, whatever it is, acts as a disrupting force to that obstruction or dissonant wave created by the obstruction. As a disrupting force, an excessive amount of it can become harmful in and of itself, which is called "overstimulation" in the acupuncture field. In effect, the force applied in "overstimulation" is so great that it overwhelms even the body's orthopathic or "upright" waveform and the patient suffers deleterious effects. Therefore, the force applied in microsystem therapy, or any other acupuncture or healing arts field, should be minimal, pursuant to maximum therapeutic effect for the patient. Therapy is often broken over multiple sessions, one reason for which being to not overwhelm the patient's orthopathic waveform, and this series of regular, repeated sessions can itself be seen as a wave, or a "superwave" or "supervibration."

The pursuit of achieving such maximal disruption to the obstruction with minimal force may be described as an 'art,' that of masterful medicine. I call it an 'art' because with the waveform model in mind, there is no one 'correct' option or solution for any disease. Hence, many dozens or hundreds of acupuncture techniques, protocols, tools, and schools exist in parallel, and all of them may be effective. The waveform model may explain why these different approaches to medicine all work, because the microsystem or other acupuncture style is only a facilitator to the body's autocorrective force to its innate waveform, which may overlap with the Western concept of the immune system. The more harmonious, consonant, or strong in amplitude the body's waveform, the easier such a process is, and that explains why some individuals may heal without the intervention of any medicine. The more unharmonious, dissonant, or weak in amplitude the body's waveform, the harder such a process is.

Returning to the field of traditional East Asian medicine, the foundational canons such as the Yellow Emperor's Internal Classic (*Huang Di Nei Jing*, dating to ca. 2nd century BCE) and the Yellow Emperor's Classic of 81 Difficulties (*Huang Di Ba Shi Yi Nan Jing*, ca. 2nd century CE) both mention the classical term *mai* 脈, usually translated in the modern day as "pulse," and usually assumed to refer to the radial pulse of the wrist. However, if one looks at the character *mai*, it is made up two radicals: the left hand radical is *rou*, for flesh or meat, and the right hand radical, according to Dr. Stephen Jackowicz, means "beat" as in a "rhythmic beat." The fundamental meaning of the character can therefore be "beat" or "rhythm in the body's flesh." This character has had variant interpretations over the centuries, including of course arterial pulse taking, but also abdominal palpation or channel palpation on the extremities, used especially in some Japanese schools of acupuncture. All of these palpatory methods led themselves to "waveform" or "vibrational" theoretical and practical approaches. The broad ambiguity of the character *mai*, "the rhythm in the body['s flesh]" can lend itself to the idea that there is a fundamental rhythm or vibration in the body that is harmonious in a state of health, dissonant in states of illness, and can therefore be influenced by externally-induced impulses or waveforms.

That leads us to the question of how music has a beneficial impact on human health. The fractal nature of the human body has been established, that some of the self-symmetry of that body is related to the golden ratio and the Fibonacci sequence has been established, that most of the intervals are based on ratios of the Fibonacci sequence has established. The theory of vibrational underpinnings of the human body and its self-healing nature has supporting arguments as mentioned above. Music may then be a resonance that, through the process of audition through the ear and engagement in the mind, engages the rational relationships or resonances within our own bodies. Since music reflects the rational and It then strengthens or helps reassert those orthopathic or "upright" relationships in the human waveform.

c2. Rhythm, Gait and the Heart

One common number that appears in the tempo of many Western songs is the number four, as in the common 4/4 time signature of four beats per measure in sheet music. In many songs of both classical and modern music, there is a superpattern of musical phrases organized in groups of four measures – sometimes eight, or even greater powers or exponents of two. Some prime examples include that of Prelude in C from *The Well Tempered Clavier* (1722) by J. S. Bach,[30] or "Bring Me To Life" (2003) by the American rock band Evanescence.[31] It can be said that the four-beat tempo (and at large, the four-measure musical "phrasing") reflects the tempo of the human gait, or more so our quadruped predecessors and zoological relatives who make up the majority of the mammalian class of the animal kingdom.

One of my first observations about Prelude in C from *The Well Tempered Clavier* is the structure of the phrasing of each measure: a doubled or repeated arpeggio of five notes. Here, J.S. Bach phrases the arpeggio in five initial rising notes, followed by three more rising notes, a repeat of the previous three "top" notes. Let us label the notes in each five-note arpeggio numbers 1, 2, 3, 4, and 5, starting from the bottom and ending at the top. Therefore, the sequence of each group of eight sixteenth-notes comprising half of each measure would be:

1 - 2 - 3 - 4 - 5 - **3** - **4** - **5**

The full structure of each measure, particular to the chord in it, from measures 1-32 and excluding the last three measured numbered 33-35 in the coda, would therefore be:

1 - 2 - 3 - 4 - 5 - **3** - **4** - **5** *
1 - 2 - 3 - 4 - 5 - **3** - **4** - **5**

Interestingly, this choice breaks what would be a mirror symmetry. It seems the natural predilection of the human mind would be to have a mirror or bilateral symmetry in the note or arpeggio progression. Where the first half of the arpeggio would ascend, the second half would most intuitively descend. The result would be as follows, making for a slightly different, if practically equivalent, Prelude in C:

1 - 2 - 3 - 4 - **5** - **4** - **3** - **2**

The resulting full measure reiterating structure:

1 - 2 - 3 - 4 - **5** - **4** - **3** - **2** *
1 - 2 - 3 - 4 - **5** - **4** - **3** - **2**

The result would be a more intuitive and congruent, if also a more pedestrian Prelude in C. The human mind finds it easier to grasp and follow the regular ascensions and descensions of both such an arpeggio, as well as counting numbers: "one two three four five four three two ONE two three four five four three two ..." as opposed to the more awkward "one two three four five **three four five**..."

Such a fact must not have been not lost on the master composer J.S. Bach, who by historical accounts was an extremely exacting man,[32] and the choice in phrasing was intentional. Either way, the result is that the latter three notes of each eight-note phrase, albeit ascending just like the first five notes, and moreover a direct translation or copy of the previous three, effect a kind of inverse or "upside-down" experience - because of the binary or mirror nature of each measure. This *ascension* becomes a kind of *descension* in the human mind - or the inverse of what the human mind would find most natural. This is a kind of paradox, maybe an experiential paradox, and perhaps one intended by Bach. The result is not unpleasant - and perhaps the break from the conventional and the expected provides a layer of novelty, tension, or suspense that engages the listener, like dissonant notes and chords are used to set the stage for "resolution" in tonal music theory. The piece at large consists of 32 consistent measures of this "inverted yet not-inverted" scalar tension that resolves in a coda with two measures that each begin with an ascending arpeggio, as before, and conclude in a *descending* scale for the very first time, leading to the final chord and resolution.

In the Prelude in C from *The Well Tempered Clavier*, the main body of the song, subtracting the "coda" or "conclusion" at the end, spans 32 measures. It is interesting to note that that number is 2 to the power of 5, where 5 is the same number of binary criteria in the Sa'am Five Element school of acupuncture, and 32 is the total number of possible combinations/results.

Examining the sequence of chords[33, 34] and listening to the song, I subjectively picked out the most notable tone for each measure. In sequence, most of these create a simple and independent kind of melody on their own, which possibly provides the overall melodic nature of the Prelude in C. I grouped this "defining" note sequence in groups of four.

1-4: E4 F4 D4 E4

5-8: A4 D4 G4 C4

9-12: C4 D3 G3 A♯4 (B♭4)

13-16: F3 A♭3 E3 F3

17-20: D3 G2 C3 G3

21-24: F2 F♯2 A♭2 G2

25-28: G2 G2 D3 E♭3

29-32: E3 D3 G2 C2

As the phrasing in the coda diverges from the above pattern, there are actually two chords in each of measures 33 and 34, with only one in the 35th:

33: C2 F4   34: B2 D5   35: C4

In a nod to the mathematical and binary regularity of the Prelude in C, there is an alternate version of the song with an additional measure by a Hamburg music director Christian Friedrich Gottlieb Schwencke (1767–1822). According to Dr. Norbert Müllemann:

> ... the piece is almost entirely subdivided into regular groups of four measures each. Though only just. One more measure – and the regularity would be perfect! So Schwencke had the feeling that the three measures, 21–23, had to be expanded into four, especially since the harmonic progression from m. 22 is in effect abrupt and audacious after m. 23. Schwencke must have thought that inadvertently this measure had gone astray in transmission and so he added it in the music text. Nevertheless, there is no doubt that Bach intended the irregularity and compression of measures 21–23; the definiteness of the source situation is not to be shaken.[35]

The medical concept of "gait," which we have now connected to the concept of musical rhythm, can connect back to the other medical concept of microsystems. Specifically, the soleus or deep muscle of the posterior calf flexors mentioned before, "the skeletal-muscle pump, peripheral heart or the sural (tricipital) pump," [16] is one of the principal muscles involved in the act of walking. Thus, the act of walking activates this reiterated kind of "heart," and it is common knowledge that walking, or exercise in general, has a clear effect or relationship with the activity of the heart proper.[36] Of course, if one stretches the imagination to all of the microsystems located in every part of the body, virtually any and all kind of exercise will activate the various constituent microsystem "hearts" in the muscle tendons and arteries. The heart was considered by many ancient cultures, including ancient Egypt,[37, 38] China, and Greece[39] to be the seat of the soul and the consciousness, the remnants of which belief remain in idiomatic usage in our English language today. The act of walking itself is widely used, both informally by laypeople and formally in traditional spiritual practices, as a kind of mental or emotional therapy. For example, there is a category of Sufi practices collectively called "Dhikr" or "Zikr" where walking or running is involved[40], and there are forms of Buddhist walking meditation[41].

One more fact worth mentioning is the number of chambers of the human heart. In mammals like ourselves and birds, there are four chambers; in amphibians and reptiles, there are three; in fish, there are two.[42] Many species of fish swim in a single planar side-to-side motion.[43] Many amphibians and reptiles also swim in a single plane, but the tailless ones, such as frogs, employ their limbs in a rhythmic one-two-resting glide motion.[44] Mammals predominantly are quadrupeds. Could the structure of the animal heart reflect their principal type of gait? Could the rhythm in music and walking, and the binary and quaternary patterns of musical phrases in music have a discrete impact on the function of the four-chambered structure of the human heart which was believed by many ancient cultures to be focal in human consciousness and emotion?

c3. Consonance, dissonance, and "stretching the human mind"

Musical scholars such as Sutcliffe and Kamien quoted above have used the words "instability," "harsh," and "unpleasant" in the case of Sutcliffe, and "unstable," "tension," "active," "harsh," "pain," "grief," and "conflict". Interestingly enough, many of these same words can be used to apply to the very nature of hard or challenging stretching, and incidentally, also the uncomfortable or sometimes painful nature of microsystem or other acupuncture therapy. Stretching itself is the introduction of an element of temporary **instability** in one's system in order to induce a change. Stretching furthermore is a challenge to the status quo of a system, which may be accustomed to being knotted and bound in its sinews and tendon structure, to change. It often is **harsh**,

unpleasant, and often elicits some **pain**. In the words of Kamien, a dissonance has "tension [which] demands an onward motion to a stable chord," and it is "moving." In the words of Sutcliffe, "dissonant intervals create a feeling of instability that needs to be resolved." These are all descriptors that fit quite neatly to the act of physical stretching, they are by their very nature **tension**, they are **motion** or **active** (to a degree, even in so-called "passive" or assisted stretching), and the person experiencing a stretch often desires to 'loosen' or 'relax' the stretch afterwards. Possibly like other physical processes such as exercise, a great part of the healing aspect of stretching occurs not during the actual stretch, during which microtrauma is being caused to the fascia and tendons, but begins once the stretch is relaxed or **resolved** and the body starts rebuilding and remodeling the newly lengthened and expanded fascia. As described by John F. Barnes, "the fascia loses its pliability" when one "experiences **emotional trauma** ... the fascia loses its pliability. It becomes tight, restricted, and a source of **tension** to the rest of the body." The patient may be caused to revisit such emotional trauma, such as **grief** and **conflict**, and hopefully release the 'trapped' tensions - at once both physical and emotional - in the fascia.

Such convergence in descriptors of a particularly evocative musical piece employing the interplay of consonance vs. dissonance and the psychosomatic effects of the act of stretching seem more than coincidence. In the absence of quantifiable evidence, I only offer my subjective experience when listening to an emotionally evocative piece of music following my preference for Romantic period classical music, for example Beethoven's "Moonlight Sonata", Piano Sonata No. 14 in C♯ minor "Quasi una fantasia." Here, Pierre Beaudry uses another interesting choice of vocabulary in describing the set of "double dissonances" at measures 52-54:

> Beethoven is forcing the listener to **stretch** his imagination to the point that he has no choice but to discover the truth of some painful experience through the second group of two dissonances, C#-D and C-C#, and then look for the reason behind his choice of the Lydian interval of the left hand to temper them. [12]

My personal experience when listening to these dissonances in this sequence, C♯ - D♮ - C♮ - C♯, is one that of stretching, yearning for something unreachable, or stretching to reach for it and failing to do so - in the notes C♯ - D♮ which have a rising diminished second interval or rising half step. The resolution, a similar rising diminished second interval or rising half step, a whole step down from the previous pair, can be emotionally interpreted as withdrawing or dropping my reach, accepting the inability to reach or attain that which was yearned for, and achieving a resolution. Consider the broader implications for this kind of emotional scenario and resolution, such as anguished patients trying to grapple with a previously unresolvable yearning for something they cannot attain or irretrievably lost. Alternatively, I feel a tight clenching or squeezing in the center of my chest, perhaps even my heart, an almost unbearable sensation, described by Beaudry as "goading pain," and then a gradual or slow release, as if the "fist" clenched around the heart let go or released. This happens twice in this sequence. This kind of phrasing in music often lends itself to playing the second repetition more softly than the first, or as a kind of "echo" - or not, depending on the interpretation of the performer. Either way, I may describe this duality and juxtaposition as a "yin and yang." Like many other classical songs, this sonata is structured in four-measure and two-measure quaternary and binary patterns integral to the body's bilateral and quadrilateral symmetry.

The membrane around the Heart in traditional East Asian Medicine, called the *xin bao*, literally "heart wrapping" or also translated as "heart envelope," "heart protector" or most often "Pericardium" is listed as a separate channel, if not a separate organ from the Heart itself. Where the *xin* Heart is considered to contain the *shen* "spirit" and shares the same word and character with the *xin* Mind in classical Chinese vocabulary, the *xin bao* Pericardium has more of an involvement with the emotions. Where the classical Chinese had a model of five interrelated emotions corresponding with the five phases or elements, which correspond to the five organs, all five of those emotions would go to a portion of the Liver called the *gan* (感 - not to be confused with 肝 *gan* for the Liver itself) where those disparate emotions would be integrated into a coherent whole and transmitted to the Heart, but first passing through its "protector," the Pericardium. The Liver and the Pericardium bear a close relationship by this connection, and they share what is called a *jue yin* or "reverting yin" relationship. *Jue* is sometimes translated as "to revert" but the word is also used to describe a "terminus" as in the terminal of a train or bus line from where the train or bus "reverts" and turns back along their route. In classical Chinese thought, yin taken to its extreme limit becomes yang, and vice versa: yang taken to its extreme limit becomes yin.

The pericardium in Western anatomy is a wrapping of fascia which does happen to serve to protect and support the beating heart. This creates a question related to my experience - could that emotional "squeezing" elicited by the sequences of dissonance and resolution in music be actually contracting and stretching the pericardium tissue around the heart? Of the pair of the Pericardium and the Liver, the Liver in traditional Chinese medical theory is thought to "govern" the tendons and sinews - which are made up of fascia. The Liver furthermore is considered to be the "storehouse" of Blood, which I have drawn several parallels to the modern concept of fascia, which has an emotional connection. Is there an appropriate connection between the "stretching" and "releasing" or "withdrawing" sensation elicited by emotionally evocative music to the "reverting" of the *jue yin* or "reverting yin" of the Liver and Pericardium?

Conclusion

This paper has explored three angles where there are connections between the elements of music and concepts of traditional East Asian Medicine. Some are not proven connections but simply correlations, and some are based on subjective human experience which is the necessary nature of music. Music, like art, is ultimately perceived by the observer.

The modern medical explanation for the sensory pathway for the experience of music, or sound in general, is through the structure of the outer and inner ear. Transmitted through the bones of the hammer, anvil, and stirrup, into the fluid of cochlea, hair cells or stereocilia perceive the vibration, which then convert them into nerve impulses to be conducted through the auditory or vestibulocochlear nerve to the brain, where it is processed or "perceived" in the temporal lobes on both hemispheres.

The traditional East Asian explanation for the sense of sound begins with the Kidneys, which send Kidney *qi* out through the ears as the senses were considered by the ancient Chinese to be active rather than passive processes, which perceive sound, which would come through one of the five orifices of the Heart through its wrapping and protector the Pericardium. The Heart is ultimately the seat of consciousness in this tradition. The perception of the sound would also go back to the Kidneys, and from there be transmitted to the *gan* or emotive affect component of the Liver. Here, the *gan* integrates the five emotions and emotive responses from the five organs and transmits them through the Pericardium to the Heart.

In either situation, the experience of the music is ultimately being perceived in the mind, whether one's definition of the mind is physically in the brain or the heart. The therapeutic effects of music, therefore, can be seen as treating the body through the mind, while the therapeutic effects of acupuncture and other body-based therapies like myofascial release, can be seen as treating the mind through the body.

References

1. Ross, Kelly L. "Mathematics & Music, after Pythagoras." Mathematics & Music, after Pythagoras, 2009, www.friesian.com/music.htm. Accessed 5 May 2017.

2. Danielou, Alain. "Chapter 1." Introduction to the Study of Musical Scales, Munshiram Manoharlal Publishers Pvt. Ltd., New Delhi, India, 2014.

3. Traynor, Robert M. "Structure of the Ear and the Leaning Tower of Pisa." Hearing International, 7 July 2015, hearinghealthmatters.org/hearinginternational/2015/structure-of-the-ear-and-the-leaning-tower-of-pisa/. Accessed 30 June 2017.

4. Peterson, Ivars. "Stepping Beyond Fibonacci Numbers." Science News, 27 Sept. 2002, www.sciencenews.org/article/stepping-beyond-fibonacci-numbers. Accessed 1 July 2017.

5. Friddle, David. "Golden Mean." The Golden Mean: Its History and Application In Music, www.davidfriddle.com/gm/Music.1/. Accessed 5 May 2017.

6. Marshall, Colin. "The Math Behind Beethoven's Music." Open Culture, 1 Apr. 2016, www.openculture.com/2016/04/the-math-behind-beethovens-music.html. Accessed 29 May 2017.

7. "Chord (Music)." Wikipedia, Wikimedia Foundation, 26 Apr. 2017, en.wikipedia.org/wiki/Chord_(music). Accessed 29 Apr. 2017.

8. "Interval (Music)." Wikipedia, Wikimedia Foundation, 25 Apr. 2017, en.wikipedia.org/wiki/Interval_(music). Accessed 29 Apr. 2017.

9. "Equal Temperament." Wikipedia, Wikimedia Foundation, 25 Apr. 2017, en.wikipedia.org/wiki/Equal_temperament. Accessed 29 Apr. 2017.

10. Sutcliffe, Tom. Consonance and Dissonance, www.harmony.org.uk/book/voice_leading/consonance_and_dissonance.htm. Accessed 1 July 2017.

11. Kamien, Roger (2008). Music: An Appreciation, sixth brief edition, student edition. Boston: McGraw-Hill Higher Education. ISBN 978-0-07-340134-8.

Retrieved from https://en.wikipedia.org/wiki/Consonance_and_dissonance 1 July 2017.

12. Beaudry, Pierre. "THE TRUTH ABOUT BEETHOVEN'S SO-CALLED "MOONLIGHT SONATA." " Pierre Beaudry's Galactic Parking Lot, 5 Aug. 2011, amatterofmind.org/Pierres_PDFs/EUROPEAN_ART/BOOK_III/2._THE_TRUTH_ABOUT_BEETHOVEN%27S_SO_CALLED_MOONLIGHT_SONATA..pdf. Accessed 1 July 2017.

13. Beethoven, Ludwig van. Moonlight Sonata (1st Mvt) Sheet Music for Piano, www.8notes.com/scores/1754.asp. Accessed 1 July 2017.

14. Song, Yi. "Hologram Theory in Acupuncture - Acupuncture Boston HolliBalance." Dr. Yi Song's Blog, www.yourhealthandlongevity.com/blog/hologram-theory-acupuncture-chinese-medicine/. Accessed 28 June 2017.

15. Hald, Connie. Connie Hald Reflexology, www.conniehald.net/history.htm. Accessed 2 July 2017.

16. Jackowicz, Stephen. Lectures 2010-2017.

17. Novella, Steven. “Iridology.” Science-Based Medicine, 21 Dec. 2011, sciencebasedmedicine.org/iridology/. Accessed 2 July 2017.

18. Meisner, Gary. “The Human Body and the Golden Ratio.” Goldennumber.net, 31 May 2012, www.goldennumber.net/human-body/. Accessed 5 May 2017.

19. “Wu Xing.” Wikipedia, Wikimedia Foundation, 5 Apr. 2017, en.wikipedia.org/wiki/Wu_Xing. Accessed 5 May 2017.

20. “Neurocranium.” Wikipedia, Wikimedia Foundation, 7 July 2017, en.wikipedia.org/wiki/Neurocranium. Accessed 8 July 2017.

21. “Soleus Muscle.” Wikipedia, Wikimedia Foundation, 7 July 2017, en.wikipedia.org/wiki/Soleus_muscle. Accessed 8 July 2017.

22. “Heart Development.” Wikipedia, Wikimedia Foundation, 1 July 2017, en.wikipedia.org/wiki/Heart_development. Accessed 8 July 2017.

23. Ahn, Chang-Beohm, et al. “Sa-Ahm Five Element Acupuncture.” Journal of Acupuncture and Meridian Studies, vol. 3, no. 3, Sept. 2010, pp. 203–213., www.sciencedirect.com/science/article/pii/S2005290110600374. Accessed 2 July 2017.

24. Gee, Henry. “The Sources of Symmetry.” Nature News, Nature Publishing Group, 6 Aug. 1998, www.nature.com/news/1998/980806/full/news980806-7.html. Accessed 2 July 2017.

25. “Gait.” Wikipedia, Wikimedia Foundation, 16 June 2017, en.wikipedia.org/wiki/Gait. Accessed 2 July 2017.

26. Goleman, Daniel, and Tara Bennett-Goleman. “RELIEVING STRESS: MIND OVER MUSCLE.” The New York Times, 27 Sept. 1986, www.nytimes.com/1986/09/28/magazine/relieving-stress-mind-over-muscle.html. Accessed 2 July 2017

27. Barnes, John F. “What Is Myofascial Release?” Myofascial Release Treatment Centers & Seminars, www.myofascialrelease.com/about/definition.aspx. Accessed 2 July 2017.

28. Barnes, John F. “What Is Fascia?” Myofascial Release Treatment Centers & Seminars, www.myofascialrelease.com/about/fascia-definition.aspx. Accessed 2 July 2017.

29. Holland, John. “The Obstruction of Sound.” The Path of Sound, 30 Jan. 2016, thepathofsound.wordpress.com/the-obstruction-of-sound/. Accessed 2 July 2017.

30. Bach, Johann Sebastian. Prelude in C from The Well Tempered Clavier, Book One. www3.northern.edu/wieland/scores/Bach/WTC1PreludeC2.pdf. Accessed 28 June 2017.

31. Ben, Moody, et al. “Evanescence - Bring Me To Life - Free Downloadable Sheet Music.” Daily Music Sheets RSS, 2003,

www.sheetmusicdownload.in/piano/sheets/1279/Evanescence_Bring_Me_To_Life.html. Accessed 28 June 2017.

32. Lehman, Bradley. "Bach's Personality." Bach Cantatas Website, 25 June 2004, www.bach-cantatas.com/Topics/Personality.htm. Accessed 28 June 2017.

33. Goeth, Philip. "JS Bach's Well Tempered Clavier Book I: Prelude and Fugue C Major, BWV 846." J.S. Bach's Well-Tempered Clavier, 2012, www.bachwelltemperedclavier.org/pf-c-major.html. Accessed 28 July 2017.

34. Motegi, Tomonari. (Revised)Bach BWV 846 Prelude Harmony & Scale Analysis (Well -Tempered Clavier, Book 1). 25 Sept. 2015, www.youtube.com/watch?v=HYgbLMe2u64. Accessed 28 June 2017.

35. Müllemann, Norbert. "On the Lookout for the Lost Measure: Bach's C-Major Prelude from the Well-Tempered Clavier I." G. Henle Verlag, G. Henle Publishers, 16 Apr. 2012, www.henle.de/blog/en/2012/04/16/on-the-lookout-for-the-lost-measure-bach%E2%80%99s-c-major-prelude-from-the-well-tempered-clavier-i/. Accessed 28 June 2017.

36. Lubans, D R, et al. "The Relationship between Heart Rate Intensity and Pedometer Step Counts in Adolescents." Journal of Sports Sciences., doi:10.1080/02640410802676687. Accessed 5 July 2017.

37. Lind, Richard E. "The Seat of Consciousness in Ancient Literature." The Seat of Consciousness in Ancient Literature, McFarland & Co., Jefferson, NC, 2007.

38. "Ancient Egyptian Concept of the Soul." Wikipedia, Wikimedia Foundation, 4 July 2017, en.wikipedia.org/wiki/Ancient_Egyptian_concept_of_the_soul. Accessed 5 July 2017.

39. "A HISTORY OF THE HEART." The History of the Heart, web.stanford.edu/class/history13/earlysciencelab/body/heartpages/heart.html. Accessed 5 June 2017.

40. Lomakin, Maxim, and Emma Wells. Zikr - Prayer on the Move in a Moscow Mosque. BBC, 26 Aug. 2016, www.bbc.com/news/av/world-europe-37191980/bbc-pop-up-zikr-prayer-on-the-move-in-a-moscow-mosque. Accessed 8 July 2017.

41. Prakhinkit, S, et al. "Effects of Buddhism Walking Meditation on Depression, Functional Fitness, and Endothelium-Dependent Vasodilation in Depressed Elderly." Journal of Alternative and Complementary Medicine, doi:10.1089/acm.2013.0205. Accessed 8 July 2017.

42. "Cardiovascular System: The Heart and Vessels of Mammals, Birds, Fish and Amphibians." Pet Education Fish, Veterinary & Aquatic Services Department, Drs. Foster & Smith, www.peteducation.com/article.cfm?c=16%2B2160&aid=2951. Accessed 8 July 2017.

43. "Fish Locomotion." Wikipedia, Wikimedia Foundation, 1 July 2017, en.wikipedia.org/wiki/Fish_locomotion. Accessed 4 July 2017.

44. "Aquatic Locomotion." Wikipedia, Wikimedia Foundation, 2 July 2017, en.wikipedia.org/wiki/Aquatic_locomotion. Accessed 4 July 2017.

Diagrams Art Sources

Figure 1:

SuperManu. File:Abdomen-head-thorax-ms.svg. Digital image. File:Abdomen-head-thorax-ms.svg - Wikimedia Commons. Scott Foresman, 7 Jan. 2008. Web. 8 July 2017. <https://commons.wikimedia.org/wiki/File:Abdomen-head-thorax-ms.svg#/media/File:Abdomen-head-thorax-en.svg>.

Modified by Raymond Hong.

Figure 2:

Created by Raymond Hong.

Figure 3:

Weisstein, Eric W. "Star Fractal." From MathWorld--A Wolfram Web Resource. Web. 8 July 2017. http://mathworld.wolfram.com/StarFractal.html <http://mathworld.wolfram.com/images/eps-gif/StarFractal_800.gif>

Figure 4:

Created by Raymond Hong.

4a. How to Do Gymnastics Jumps. Digital image. WikiHow. WikiHow, n.d. Web. 8 July 2017. <http://www.wikihow.com/Do-Gymnastics-Jumps>.

4b. Fageot, Christine. "FENG SHUI - and Your Property." FENG SHUI and Your Property. N.p., n.d. Web. 8 July 2017. <http://ab-real-estate.com/newsletters/nl43/feng-shui.php>.

4c. Winkler, Mary. Create Gender and Orientation Symbols With Basic Shapes in Illustrator. Digital image. Design & Illustration Envato Tuts. N.p., 16 Oct. 2013. Web. 8 July 2017. <https://design.tutsplus.com/articles/create-gender-and-orientation-symbols-with-basic-shapes-in-illustrator--vector-17634>.

Figure 5:

Created by Raymond Hong.

5a. The Winking Emoticon. Digital image. King Mag. N.p., 31 Jan. 2008. Web. 8 July 2017. <http://king-mag.com/the-winking-emoticon/>..

5b. Bikmullina, Olga. Art Tuts. Digital image. Inkscape Tutorials | How to Draw a Crown. N.p., n.d. Web. 8 July 2017. <http://tuts.ahninniah.graphics/inkscape-tutorials-how-to-draw-a-crown/>.

Contributors

David O. Akombo, Ph.D. is Associate Professor and Graduate Fellow for the College of Liberal Arts at Jackson State University. Before joining Jackson State University, Dr. Akombo was a Diversity Fellow at Weber State University in Utah. Intrigued by the peculiarities of music and its interdisciplinary confluence with medical and psychological practices, Dr. Akombo has studied and researched the effects of music on children and young adults both in schools and in hospitals. As scholar and performer, Dr. Akombo balances his research interests amicably as a music educator, ethnomusicologist, composer and singer/drummer, having worked in Africa and Southeast Asia where he studied with Balinese artists. His books, *Music and Healing Across Cultures* (Ames, Iowa: Culicidae Press, 2006) and *Music and Medicine: Connections Found* (Seaburn, 2009) unfold the mechanics of the relationships between music, culture, physiological and spiritual states in people from the anthropological and epistemological inquiry regarding music healing and the cosmos. These books show the power of music and its ability to promote mind/body coordination. Dr. Akombo has worked jointly with the University of Florida's Center for Arts in Healthcare Research and Education (CAHRE) with which he collaborated to establish the Arts in Medicine (AIM) program in Nairobi, Kenya. Located at Mater Hospital and funded by the Legislature and State of Florida Center for Cultural Affairs in Tallahassee, this is the first known Arts in Medicine program in Africa. Dr. Akombo continues to be active in music education research, and interdisciplinary scholarship having conducted clinics in several countries. He is a 2017 recipient of the NSF Research Training Scholarship. He is member of the Music Educators National Conference, The national Social Science Association and co-founder and member of the Executive Board of NSSA and ISQRMM.
dakombo@isqrmm.com

L. Elise Carter is a graduate student in Music Therapy at Montclair State University in Montclair, New Jersey. Prior and in addition to going back to school, Ms. Carter has a career as a musician which encompasses many things. She is a co-founder of the New Jersey Flute Society, a member of the chamber ensemble Uptown Flutes, and adjunct flute professor at Drew University. She is a teacher and freelancer in the NY metropolitan area. She is an award-winning composer published by Theodore Presser and Art-of-Sound Music. She is also the music director at Calvary Lutheran Church in Allendale, New Jersey where she plays organ, piano and conducts the choir. Ms. Carter is the founder of the non-profit Carter Glennon Center for Arts and Nature. www.carter-glennon.org
longnote@msn.com

Joe Carucci, DMA was raised in Clifton Park, New York. Saxophonist, composer, and educator, Dr. Carucci received a Bachelor of Music from the Eastman School of Music and a Master of Music and Doctor of Musical Arts from University of Kentucky. He has studied saxophone with Dr. Ramon Ricker, Professor Miles Osland, and the late Mr. Nick Brignola. Joe has performed and recorded throughout the United States, Canada, Japan, and China. Many of his compositions are available through Walrus Publishing, Jazz Lines Publications and have been used on Pioneer Public Television for various programs. Joe is the Director of Music Industry at Eastern Kentucky University. Prior to his appointment at EKU, he was on faculty at Lincoln Memorial University, University of Minnesota-Morris, and Transylvania University. He has also taught classroom and instrumental music in California, Kentucky, Massachusetts, and New York.
joseph.carucci@eku.edu

Derrick Crow, Ph.D. holds a Ph.D. in Education with a concentration in Arts Integration and Curricular Theory. He also holds a Bachelor's of Music Education from Eastern Illinois University, and a Master's of Music from Southern Illinois University Carbondale. His research interests include music medicine, aesthetic education, and arts integration. Through research and pragmatic experiences, he is striving to create educational growth that encourages creative and intellectual exploration. He has had the privilege to work as a music educator at the primary, secondary and collegiate levels, has worked as a student teacher supervisor for the Master's of Arts in Teaching Program, and as an adjunct lecturer at Southern Illinois University Carbondale. He is currently practicing his skills as an educator and musician by teaching young musicians in Danville School District 118.
crow.da19@gmail.com

Dena Gold, M.S.Ac., MSCH is a licensed acupuncturist and massage therapist working in private practice for over five years. She has Masters degree in Acupuncture from Tristate College of Acupuncture and Masters degree in Chinese Herbology from University of Bridgeport. Through her study of both modern and classical Chinese, she is committed to expanding her understanding of Chinese medicine via primary sources. Her deep interest in music has always compelled her to seek to understand the music of other cultures. As an aerial dancer Dena has collaborated with New York based musicians including Gabriel Marin, Dan Kurfirst, Tom Swafford, and Joshua Torres.
denalgold@gmail.com

Raymond Hong, L.Ac., MSAOM is a licensed acupuncturist and herbal program graduate of New York College of Health Professions, class of 2013. He also studied with the Institute for Classical Asian Studies for nine years. He currently practices traditional East Asian medicine. His undergraduate major was Middle East Asian Languages and Cultures at Columbia University in the City of New York. Raymond has been a student of the cello since the age of seven, participating in student orchestral, chamber, and solo performances at the Center for Preparatory Studies in Music at Queens College, Phillips Andover Academy, Point CounterPoint chamber music camp, and the Children's Orchestra Society. He also has participated in student programs in Korean folk music on drums and the zither.
raymondhongacu@gmail.com

Carla Rose Kelly is a life-long musician, multi-instrumentalist, award-winning pianist, composer, and Master Healer with Music from London, Canada. She has three post-secondary degrees, in addition to her Conservatory training, and Masters in Healing Arts. While teaching Royal Conservatory piano & theory for 25 years, she became involved in Music & Medicine research. She has had the privilege of speaking internationally about her published research, hosting workshops and retreats in Japan, England, Norway, Canada and USA.

She developed a holistic music healing program with methodology and SACRED SCALE® □ patented musical instruments specifically designed for therapeutic use. She teaches classes and conducts group and private sessions using this music for healing.
She is also a playwright, composer and producer of the 2018 *Let the Music Come Alive,* a new Canadian musical telling the tale of how a young girl uses music to help her father's heart.
info@carlarosekelly.com

Roy Kennedy, Ph.D., MT-BC, LPMT is an associate professor of music therapy at the University of Georgia, USA. While previously providing music therapy services to various populations, his primary research interests include guitar pedagogy for music therapists, music therapy for English-as-Second Language learners, adults with substance abuse problems, and skill generalization of students with developmental disabilities. Dr. Kennedy has presented his work in the USA, Brazil, South Korea, and Costa Rica. He continues having an interest in the international development of the profession.
rkennedy@uga.edu

Eve Kodiak, M.M. specializes in the interplay of developmental movement, music, energy, and emptiness. A licensed Brain Gym consultant, she has a healing and educational practice at The Lydian Center in Cambridge, MA. A Kundalini Yoga Teacher and author of *Rappin' on the Reflexes: A Practical Guide to Infant Reflexes,* Kodiak presents internationally and is a recording artist. Her latest work has been a collaborative interdisciplinary project with the company *Colors In Motion,* creating scores for fine art digital media to create calm in public spaces.
eve@evekodiak.com

Chyi June Khoo, MM, MA, MT-BC is a music therapist hailing from Malaysia. She acquired her Master's degree in Music Therapy from Montclair State University. Her academic background comprises an undergraduate degree in piano performance from China Conservatory and also a Master's degree in piano performance from Rowan University. Her clinical work in music therapy includes work at Seasons Hospice and Palliative Care, Elizabeth Seton Pediatric Center, Morristown Medical Center and Monmouth Medical Center. Her professional interests are music therapy and the Disorder of Consciousness, and the effect of music on brain activity. Chyi June is the recipient of the David Ott Scholarship for Outstanding Music Therapy Intern Award in 2016.
khooc1@montclair.edu

Kristie Lipford, Ph.D. is a NIH Disparities Research Trainee and a junior faculty member at Jackson State University where she divides her time between teaching and research. In her teaching role, Dr. Lipford oversees and plans curriculum for the Research Methods course in the Bachelor of Social Work Program. In her research role, she broadly examines health behaviors. She is currently completing a manuscript that examines the effect of psycho-social stressors on physical activity, dietary habits, and sleep patterns using Jackson Heart Study data. Her most recent focus though is integrative medicine. At the present, she is preparing a NIH Career Development Award to undergo a review with the National Center for Complementary and Integrative Health. Dr. Lipford is interested in analyzing locally, nationally, and globally available datasets to examine complementary and alternative medicine use among ethnic groups and minority populations.
kristie.lipford@emory.edu

Julia MacKinley is a graduate of the Psychology Program at the University of Western Ontario and pursued a graduate degree at the University of Edinburgh in Human Cognitive Neuropsychology. In addition, she is a member of the Canadian Psychological Association. Her research started with an interest in posttraumatic stress disorder and how individuals process information about themselves. This research later developed into an interest in the uses of yoga, meditation and music as forms of healing, particularly in the context of psychiatric illness. Currently, Julia works as a psychometrist in adolescent and adult mental health with a focus on rehabilitation and recovery.
julia.mackinley@gmail.com

Wendy Thompson, Ph.D. (***This contribution work is published posthumously.) was an Associate Professor of the Graduate Social Work Program at Jackson State University. She previously served as an Associate Professor of Social Work at Andrews University for three years where she taught research methods, human behavior, critical thinking, dealing with difficult people, families and children at risk and social policy, both at undergraduate and graduate levels. Currently, at Jackson State University she is teaching social work ethics, psychopathology, social work practice, research methods and advanced direct practice seminar. Wendy worked in a variety of settings such as community head- start, public school, drug and alcohol rehabilitation facility and department of health. Thompson has authored three peer-reviewed articles, presented at professional conferences and edited articles for the International Journal of Environmental Research and Public Health and Drug Abuse Issue of The Journal of Adventist Education. Dr. Thompson's research focus encompasses sickle cell disease, drug and alcohol abuse, domestic violence and disparities in health care.